Captain Jeremy Burfoot is a retired Qantas pilot with 23,000 hours of flying between over 100 countries under his belt. He worked for Qantas for thirty-six years, Japan Airlines for three years and was a navigator in the Royal New Zealand Air Force for three years. He has three world records on a jet ski (including 24-hour longest distance travelled), and has raised $500,000 for melanoma and prostate cancer. He has represented New Zealand cycling at world masters track championships multiple times.

Today, he delivers workshops in overcoming the fear of flying, and is an author and an adventurer. You can find out more and connect with Jeremy at jeremyburfoot.com

THE
SECRET LIFE
OF FLYING

CAPTAIN JEREMY BURFOOT

MACMILLAN
Pan Macmillan Australia

Pan Macmillan acknowledges the Traditional Custodians of Country throughout Australia and their connections to lands, waters and communities. We pay our respect to Elders past and present and extend that respect to all Aboriginal and Torres Strait Islander peoples today. We honour more than sixty thousand years of storytelling, art and culture.

Some of the people in this book have had their names changed to protect their identities.

First published 2024 in Macmillan by Pan Macmillan Australia Pty Ltd
1 Market Street, Sydney, New South Wales, Australia, 2000

A catalogue record for this book is available from the National Library of Australia

Typeset in 12/17 pt Fairfield by Midland Typesetters, Australia
Printed by IVE

Image on page 4 (upper) by AlexandraDaryl/Adobe Stock Images; page 4 (lower) adapted from Michael Paetzold/Wikimedia, CC BY-SA-4.0; page 119 by BooblGum/Adobe Stock Images.

The paper in this book is FSC® certified. FSC® promotes environmentally responsible, socially beneficial and economically viable management of the world's forests.

THE FLIGHT PLAN

'The Wright brothers created the single greatest cultural force since the invention of writing. The airplane became the first World Wide Web, bringing people, languages, ideas, and values together.'

Bill Gates

'If you can walk away from a landing, it's a good landing. If you can use the aircraft the next day, it's an outstanding landing.'

Chuck Yeager

INTRODUCTION

My flying career wasn't supposed to end the way it did. I expected it would end at retirement age, with my last landing being rubbish because I tried too hard, and I'd have to live with that for the rest of my life. The landing would be followed by the age-old tradition of the airport fire engines spraying the aircraft with water. Then I'd taxi into the terminal. There, I'd be met by my boss at the gate, who would shake my hand and steal my gold watch.

But thanks to Covid, it ended with a whimper rather than a bang, and I didn't even know I was making my last landing. I got back home from London just like I always have, but a few days later, they shut the operation down. I never flew again. But there is a silver lining, for you anyway, because now that I don't work for the mob, I can tell you what really happens behind the scenes in an airline operation without being 'invited' to seek alternative employment.

You'll get to read all about flying from an insider's perspective. I'll show you exactly what goes into getting you up in the air and down again safely, which takes many complex systems, and many different people with essential roles, all working in unison. It's remarkable, and you'll come away as amazed about the reality of it as I still am (with over 23,000 hours of piloting aircraft behind me!). Flying is about complex technology and systems, but it's also about people, on the ground as well as in the air. And where there are people, there are different complexities involved – and some very good stories. The next time you board a plane and strap yourself into your seat, I'd like you to be thinking of the aviation history, in pioneering, in engineering, in man hours and lives that it has taken to get us to this point in time, when it feels so natural to entrust our lives to complete strangers and cross the world when we want to.

I always found flying super exciting and mentally stimulating. The adrenaline buzz never got old. Every flight was a new adventure even if it was to the same destination. As the famous Greek philosopher, Heraclitus said, 'No man ever steps in the same river twice, for it's not the same river and he's not the same man.'

My mission or 'flight plan' for this book is to present everything in terms of a flight from start to finish so that it helps put everything in context, thus making it seem more logical. It's how we pilots think.

I also like to see the lighter side of everything and tell it how it is. It's a more relaxed way to be. My goal with this book is for you to finish knowing a whole lot more than you did at the start and having been entertained as well along the way.

SOME USEFUL INFORMATION

Other than balloons, aircraft all fly using the laws of aerodynamics. Since the theme of this book is aviation, it seems like
an excellent idea to talk about these basic laws now. They will
go a long way towards making things more straightforward as
we talk about the development of aircraft and the hard lessons
learned along the way. The study of aerodynamics is a highly
complicated field if you choose to dig into it, but we'll keep it
simple here.

Aerodynamics is the study of the movement of air around
objects. In aviation, the four laws of aerodynamics underpin
everything from the design of aircraft to the principles of flight,
and how aeroplanes are able to take off, fly, and land.

The first law of aerodynamics is the law of lift. This law states
that an object (in this case, an aeroplane) moving through a fluid
(air) will experience a force perpendicular to the direction of
motion. This force is known as lift and comes from the shape
of the wing. The curved wing, known as the air foil, creates a
difference in air pressure between the top and bottom surfaces

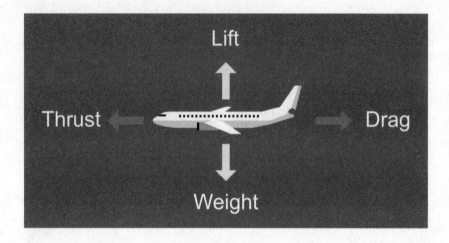

of the wing, which results in a net upward force. Bernoulli's principle explains how the air has a longer distance to flow over the curved upper surface of the air foil, so it speeds up, reducing the pressure on the top of the air foil as it does. If you hold your hand out the window of a car at a slight angle to the oncoming airflow, a high-pressure area will form under your hand and lift your arm. Don't try this in traffic.

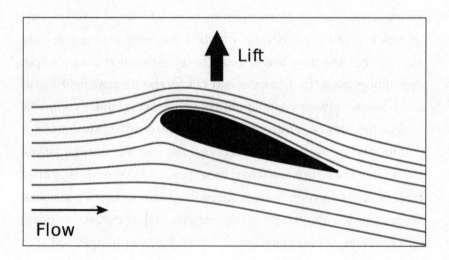

The second law of aerodynamics is the law of weight. This law states that all objects have weight and that this weight must be overcome for an object to achieve flight. For an aeroplane, this weight includes how heavy the aircraft is as well as its fuel, passengers, and cargo. For the aircraft to leave the ground, lift must be greater than the weight.

The third law of aerodynamics is the law of thrust. This law states that in order to achieve and maintain flight, an aeroplane must have a forward force known as thrust. A jet engine or propeller typically provides this forward force, which propels the aeroplane forward through the air. This creates the airflow over the wing and hence the lift.

The fourth law of aerodynamics is the law of drag. This law states that all objects moving through a fluid will experience resistance, known as drag. When you walk into a strong wind, it's drag that's holding you back. For an aeroplane, this drag is caused by the friction between the aircraft and the air around it. Therefore, an aeroplane must be as streamlined as possible to overcome this drag.

Stalling is also worth mentioning here, because this is the point at which aircraft stop flying, which is not ideal and generally to be avoided. When a plane flies, the air around the wings is at different pressures. As mentioned above, the air passing over the wing has a lower pressure than the air passing under the wing, generating lift and 'pushing' the aircraft upwards.

However, when an aircraft increases its 'angle of attack', known as the angle at which the wings face the oncoming air, a separated flow of air is created behind the wings where the two air pressures mix. At a certain point, the separated flow reaches a critical mass that stops lift generation. Without lift, the aircraft will start to fall no matter how powerful the engines are or how fast it flies. The point where an aircraft wing reaches stalling conditions by raising the nose of the plane is called the critical angle of attack.

So all aircraft, no matter what type, operate using these principles. It's unavoidable. When I was swapping from the Boeing 747 to the Airbus A380, a few people asked me if I was worried about how differently the two types flew and how the flight control systems were different. Would I cope, they asked. My answer was that both are subject to the same aerodynamic laws. Neither manufacturer has worked out how to avoid them yet. I was not concerned.

As well as how lift works, it seems appropriate to talk about some of the conventions of aviation here that will pop up later in the book.

We mark locations using latitude and longitude. Degrees of latitude start at the equator and go to 90 degrees north at the North Pole and 90 degrees south at the South Pole. Degrees of longitude are 'vertical', from pole to pole, starting at the Greenwich meridian (a line that goes through Greenwich, England) and going east to 180 degrees east and west to 180 degrees west.

In aviation, speeds are measured in knots (kt). Knots are nautical miles per hour. A nautical mile (nm) is 1,852 metres and corresponds to a minute or 1/60th of a degree of latitude. Aviation shares maritime measurements because the link with latitude historically made navigation easier. In this book I'll often give the equivalent speed in kilometres per hour (kph) as well. Altitude is measured in feet (ft) in most countries, metres (m) in a few. Later in the book we'll discuss how altitude measurements change as we fly. Weight in most countries is measured in kilograms (kg) or tonnes (t), though the standard is to use pounds (lb) instead of kilograms in measurements like engine thrust, and pounds per square inch (psi) for pressure. Temperature is normally in degrees Celsius (°C).

Lastly, the international language for aviation communications is English.

So, now we're all in the know about some of the basics, let's get ready to fly.

ONE

PRE-FLIGHT

HURRY UP AND WAIT.

AT THE DEPARTURE GATE

Let's start at the departure gate. You've navigated check-in – and border controls and duty-free shopping if you're flying internationally – and have put your clothes back on after being forced to get almost naked at security (and if you were clever, you will have upgraded your phone and shoes here too). Now you're in that state of suspended animation, waiting for the boarding announcement that will start your journey.

If you look out the window, you will see feverish activity going on around your aircraft. This is all done by ground workers, and it's not very glamorous. Inside the terminal, you only see the more glamorous side of aviation, such as the cafes, overpriced boutiques, pilots strutting around like peacocks and flight attendants done up to the nines. There's a whole other world going on outside the terminal.

Every aircraft you see at the airport requires a ground crew to maintain it and get it back into the air. On one of the large double-decker four-engine planes, such as the Boeing 747 or the Airbus A380 – ranked in rough order of pay scales, from low at the top end to sweatshop at the bottom – there are aircraft engineers (2), refuelers (1), ramp supervisors (2), security people (3), bus drivers (it depends), people who wreck and lose your bags (5), catering crew (8), people who empty the poo from your aircraft (1) and aircraft cleaning crew (12).

What do I, a former pilot who used to strut around the airport like a peacock, know about ground crew, you ask? Besides actually talking to them throughout my career, for a while after Covid grounded my career as a pilot, I took a refuelling job at Auckland Airport to get an idea of how the ground operations work for a consulting company I had started up.

About the time I got the refuelling job, I was also offered a job sitting in a car at the airport all day trying to catch other ground workers out for breaking the rules. The girl who phoned me said, 'With your experience in aviation as an A380 captain, you'd fit in well with this job, and we'd like to reward you with minimum wage.' I'm not sure why a job as an A380 captain made me so qualified, but I wasn't tempted and politely declined.

One thing I noticed in my time on the tarmac is that a lot of the workers, many of whom are new due to the recruitment boom that happened post-pandemic, don't have the right level of experience or safety nous, and as a result, many shonky and dangerous things go on that will cost the airlines, over time, with injuries and flight delays.

But here's the thing. These ground workers, whom you are mostly unaware of, have families and bills to pay just like you, but most are on minimum wages. That's so you can fly somewhere for a lot less than you should be paying, or so airline executives can be paid their bonuses (you choose). These are real people doing good and sometimes dangerous work, mainly for no thanks.

But as Winston, a poo tanker guy, said to me once, 'At least it's a job in aviation, and my parents are very proud of me.' Winston's parents don't know exactly what he does. All they know is that he is an evacuations specialist, and Winston is the first there if the shit ever hits the fan.

The enthusiasm on the tarmac is infectious, which shows the calibre of the people there. The ground workers I've had dealings with, in my career, are a fun and happy bunch who constantly work with a smile. Of course, they could get almost as much money on a benefit, but they aren't and kudos to them for that.

When an aircraft arrives at the gate for a short turnaround, it's like a tyre change and refuel for a Formula 1 car. It's controlled mayhem. The only way it gets done on time is for everyone to charge around at breakneck speed and do things that aren't always safe. I discussed this with my refuelling boss and asked if the process was ever audited. He said, 'Yes. But management, who have KPIs for good audit results, tell everyone in advance when it will happen, so they never fail.'

To be more specific about the dangers of working on the ground, we are talking about multiple vehicles driving around at speed in all weather and often in the dark. Apart from vehicles, there are all sorts of loading equipment being used to move and lift heavy containers. Being run over or having body parts crushed is not uncommon. Sadly, even being sucked into running jet engines is not unheard of.

I know how much you all hate flight delays, so this needs to be fixed, but I'm not sure what the answer is. If I was running the place, I'd pay them a bit more, encourage loyalty and make them stakeholders in the safety culture of the company. That would be a good start, at least.

In other news from the ground operation, I have seen why your bag sometimes doesn't turn up and other times turns up wet. Some smaller aircraft, like the Boeing 737, load all bags into a bulk baggage hold by hand, off a trolley which is towed from the terminal stacked to the sky with your bags. Frequently they fall off en route. The tow tractor doesn't have rear-vision mirrors, so unless the driver is Linda Blair, they won't see your bag fall off. If it rains, all the bags on top of the pile get wet. Larger aircraft tend to have bags loaded in semi-waterproof containers, which is a much better system.

Around 0.5 per cent of all bags get lost on average worldwide. That means 1 in 200. On a B737, that's one bag on each flight. Three on a B747. That's 25 million bags worldwide per year. It is said that the rings of Saturn are made up of lost airline bags, which explains why the rings are so visible. The good news is that the percentage of lost bags has halved in the last 15 years. Apparently, 16 per cent of misplaced bags are a result of the wrong tag being put on a bag, so that's something you can check yourself before you bid your bag farewell at the check-in counter.

Since I did the refuelling gig for a while, it would be remiss of me not to explain how that works. In most big airports, fuel is pumped under pressure through a large underground pipe to hydrants in the tarmac. The refueller drives his truck under the wing, gets out and runs all sorts of wires as if the aircraft is getting an ECG. He plugs into the hydrant at one end and the refuelling points under the aircraft wing at the other end, then allows the pressure from the hydrant to push fuel into the aircraft. The truck and its systems are really only a metering unit. The system is capable of up to 4,000 litres per minute if the aircraft will accept that much. It's pretty impressive.

Pets get a bum deal flying by air. The environment around the aircraft's outside is as noisy as a mosh pit at a Nirvana concert. Workers are supposed to wear earmuffs and sometimes earplugs as well. But pets are often left on the tarmac in this noise for long periods of time. They are not even offered earmuffs; if they were, they probably wouldn't wear them because they'd feel self-conscious. So, if your cat ignores you after a flight, don't take it personally. It's just deaf.

Meanwhile, as you decide whether to run back for another coffee, in some dingy airport office, the cabin crew and the pilots have been briefing for the flight from around 75 minutes before departure.

The pilots will start with greetings and idle chat and work on forming a team. This team forming is one of the most wonderful things about large airlines, and pilots should be able to do it well. The team, at a minimum, will consist of a captain and a first officer. The captain is effectively the CEO of the entire operation, and the first officer is his deputy. If the flight is a long one, over 14 hours, a relief pilot, or second officer, will be added to the crew. Ultra-long flights over 16 hours require two second officers so that all four pilots get adequate rest.

These two to four pilots, who might never have met before, have an hour to form a cooperative team and get a huge jet into the air. And, of course, this team building can occasionally be tricky because of personalities. The odd grumpy or unpopular captain may struggle with it, but these captains are normally rescued from themselves by the sheer professionalism of the other pilots.

As a captain, you lead the team-building exercise and how well you do reflects on the performance of the operation. My personal favourite way was to start with a derogatory comment about the airline CEO, who was not well liked by flying staff. This always worked, and it became my go-to team-building method. Then, with team building ticked off, we'd move on to other things.

The pilots will have received a flight plan, weather reports and any other notices pertinent to the flight on their iPad. They then go into a huddle and decide how much fuel to take. The flight plan will have a minimum fuel figure on it, but it's up to the pilots to determine how much extra fuel to take, if any. In deciding this figure, the pilots consider the weather, plus possible delays in the flight due to congestion. Aircraft use more fuel if they fly below their optimum altitude, so it's a contest of sorts to beat the other airlines to the higher levels.

Carrying extra fuel costs money because of the extra weight. If you just fill the tanks, the airline would soon go bust, so management 'encourage' the pilots to take the minimum fuel. There is always a certain amount of antagonism over this, as the overriding issue for pilots is not running out or even getting close to that. Running out would be stressful, start quietly and end noisily.

The planning process and the overall operation of the flight is a finely tuned balancing act between safety and commercial reality. The number-one responsibility is safety, but commercial realities need to be considered, and professional pride dictates that you combine these two considerations to the best of your ability. To do this well requires years of experience. Management monitors the carriage of excess fuel and the use of inefficient operating procedures. The odd pilot who consistently fails to produce will be counselled, hopefully in a positive way.

Sometimes even full tanks are not enough fuel. The weight of the fuel is what gives you your energy. In hot climates each litre of fuel weighs less than in cold climates, because the volume of fuel expands in warmer air for the same weight. This is known as the specific gravity of the fuel. In hot conditions, the tanks will sometimes fill up volumetrically before you get the required weight of fuel into them.

With an A380 load of 320,000 litres, at 40°C there is a 3.7 per cent reduction in the weight of fuel that can be loaded compared to at 0°C. That's 9,400 kg less fuel, which is 40 minutes of flying in an A380. Sometimes 40 minutes of fuel can mean the difference between making the destination legally or not.

Weather forecasts are improving in advanced countries thanks to computer modelling and historical data, but you can never be sure. I always knew the weather would be an issue after driving past the local weatherman's house while in the air force. It had just started pouring with rain, and he and his mates were hurriedly packing up their barbecue. Accurate forecasting was a concern then and it remains so, to this day.

Eventually, the pilots decide on a fuel figure and advise operations, who send that figure to the refueller sometime before Christmas. Operations will then get grumpy when the refueller holds up the aircraft because he's late finishing off the refuelling. Refuelling some larger aircraft can take over an hour, so an early fuel figure is helpful.

As mentioned, an Airbus A380 can carry up to 320,000 litres of jet fuel. That's a weight of 256 t. It goes into tanks in the wings and a tank in the tail area. On the ground, that sort of weight hanging off the wing roots and undercarriage is quite stressful on the aircraft. Once the aircraft is in the air, that stress is reduced as it all sits in a lifting surface. Aircraft prefer to be in the air.

Some airlines won't refuel with passengers on board. Most will, though, and there are several requirements for doing this, such as having an aerobridge attached and a door open.

After the planning and briefing are complete, the crew leave the dingy office, stroll past you at the boarding gate, board the plane and begin their pre-flight checks in preparation for welcoming you on board.

For the cabin crew, this involves checking all the safety and communications equipment and packing catering into the right place for later. This is not always as simple and straightforward as it may seem. For example, galley checks are important, especially the Passenger Special Meal List – Vegan, Vegetarian, No Gluten etc. I can remember, out of Singapore, seeing the rather long (almost a toilet roll) list of special meals that the cabin crew would have to deal with. The cabin crew have to be very organised in getting these meals checked off by opening all those meal carts and making sure the meals (with the passenger's name on it) match the list they have!

For the pilots, their pre-flight duties include checking all the systems are working and configuring them correctly. These systems include hydraulics, electrical, pressurisation, flight controls, navigation, software, fuel, engines and many more. A lot of this checking is just ensuring all the switches are in the correct position, as warning systems will alert pilots to any faults or incorrect configurations.

When this is done, the pilots will load the flight plan into the computer and brief each other on their intentions. At some point, one of the pilots will go downstairs and have a wander around to see that nothing significant has been stolen, that nothing is falling off the aircraft, and there is still air in the tyres. It's a myth that they go down to kick the tyres. At 220 psi tyre pressure, this would not prove much, but it would hurt a lot.

At some point during the pre-flight, the cabin crew will also deliver a designated crew snack to the cockpit. The idea of this crew snack is to raise the blood sugar of the crew prior to the intense departure phase. Cabin crew normally get one as well. This is a really good idea if done properly, but it seldom is. Because accountants are involved at every point in an airline's operation, these snacks are generally missing one important ingredient: food that you would want to eat. So instead of Liquorice Allsorts and Big Macs, they are generally full of unidentifiable green stuff and dry bread that was bought on sale after it passed its use-by date, in the year prior. In a high percentage of cases, these crew snacks end up as pig food at the destination. They are mostly ignored and everyone gets on with the job of preparing for departure.

BOARDING

At some point, about half an hour before departure time, you will be invited to board the aircraft. If you are in business, first class, or high up on the airline's points scheme, you can go first. Then it's usually done by seat row, as they like filling the rear of the aircraft first. Unfortunately, some people like to ignore boarding instructions, either because they don't speak the language or because they have a lot of carry-on bags and want to take advantage of space in the overhead lockers. The only reason ground staff don't carry sticks to beat these people with is that beating passengers is discouraged.

Ideally, boarding would be done at the speed of a Benny Hill chase and with the Benny Hill chase music playing, but sadly it's more like the last steps of a death-row prisoner towards the electric chair. Everyone agrees that this should be fixed, but no one has figured out how. The Foo Fighters put out a music video called 'Learn to Fly' which runs through a boarding and a flight, and is worth a watch if you have a sense of humour. But while it's easy to pick holes in the process, there is still no real answer. So, there's a challenge for you. Perhaps reducing the number of seats, much like musical chairs, might create some urgency. It's worth a crack.

If you think about it, though, what is the hurry unless there's a chance of departing early? Why else would you want to spend more time on an aircraft than you had to? I like to leave boarding as late as possible unless I'm in business class.

If you do board early in the process, there are some downsides. One is that if you are sitting on the aisle, everyone will bump into you with their bags and butts. You are also subjected to the lottery of watching everyone walking down the aisle towards you and wondering if they will be sitting next to you. If the person is attractive and the gender you are attracted to, you will be thinking, 'Yes! Please sit next to me and be playful, naughty, rich and seriously attracted to me.' If the person heading your way weighs north of 150 kg and obviously can't shower properly without assistance, you will be using all your powers of telepathy to direct them to a different seat. Crying babies and badly behaved children can also be a nightmare.

The subject of overweight people on aircraft has been in the news of late. People who struggle to fit on one seat have been complaining about how unfair it is that they are, in some cases, being forced to buy two seats. It's hard to argue for or against this without offending someone. To make seats bigger or force airlines to give extra seats to these people for free would reflect on the bottom line and end up costing everybody more. Skinny people could argue that they were then being discriminated against by having to pay more for their seats. A solution to this won't be easy to find.

If you have been waiting for boarding time in an airport bar, you need to be aware that it's not okay to turn up at the boarding gate drunk. Have you ever turned up late to a party and been sober while everyone else was drunk? If so, you will know what I mean when I say it is always obvious when someone else is drunk. Sadly, in the past, letting drunk people on board because they promised to behave has usually not ended well, so the crew have zero patience for this these days. I remember letting a drunk passenger board early in my career. It was on the understanding that he wouldn't be served alcohol, but he had his own bottle in his bag and continued to drink on the sly from that. Eventually he got out of control and violent and had to be restrained with handcuffs.

As a captain, if the customer service manager came to me before departure with a report of a drunk passenger, I would always order the offending passenger off the aircraft. In many countries, the law says, 'An intoxicated person may not enter an aircraft.'

In other news, being down the back means you will be last off at the end of the flight and probably sitting next to the toilets as well. But, statistically, you are slightly more likely to survive a crash seated down the back. This is logical, considering that aircraft do not fly into mountains backwards, and you nearly always see a tailfin in the wreckage. According to an FAA report on accidents and deaths between 1985 and 2020, the middle of the plane is the most dangerous place to sit. It's no coincidence that this is where most of the fuel is. Sitting in the middle third of the cabin put the risk of dying at 39 per cent, while being in the front third marginally reduced that risk, to 38 per cent. Taking a seat in the rear third of the cabin, however, brought the risk of death down to 32 per cent. You also need to consider that being last off can put you behind everyone in the customs queue too. That's not ideal, but it's still not as inconvenient as death.

If the aircraft is half empty and you want to move away from your current seating companions before departure, you can't. This is because the aircraft will have been loaded to keep the centre of gravity within limits for take-off. So, when your cabin crew says you can't move, accept it and make plans to extricate yourself and dash for better seats as soon as the seatbelt sign goes off. This doesn't apply to moving sideways, which doesn't affect the centre of gravity. But no, you can't move sideways up or down the aisle.

And now, with the aircraft nearly full, you could be forgiven for assuming that things are going along swimmingly, and it won't be long before aviation is being committed at a furious rate. You will possibly be thanking the Lord that the seat next to you is empty as well. Maybe it is time to buy that lottery ticket after all. In the cockpit, the pilots are thinking along similar lines and the captain has just asked the second officer to call the customer service manager and ask if we are close to closing the doors.

The response everyone is hoping for is, 'Everyone on board, boss, just waiting for the door to be closed.' But way too often it's, 'Just waiting for one. The ground staff are looking for them now.'

I can assure you that this is as annoying for the pilots as it is for the passengers waiting to go. People who turn up late for flights are either arrogant or as dumb as a post. There are no in-betweens. In my opinion, both types should be left behind: the arrogant to teach them a lesson and the dumb because they are too stupid to let on the flight. They may cause more significant problems, such as lighting a fire in the aisle to cook breakfast.

There was one particular man who was a third dan black belt in turning up late due to arrogance. In 2017 we were waiting to depart Tokyo's Haneda Airport in the evening heading for Sydney. We were missing one passenger. From the cockpit we could see into the departure gate through large glass windows. About ten minutes after the scheduled departure time, we saw the passenger wander in at a relaxed pace.

He boarded and we departed, thinking this was the end of it. Once we arrived in Sydney, I picked up my bag and checked in for my flight to Auckland to commute home. I boarded that flight, and at departure time the captain announced we were waiting for one passenger. Ten minutes later, the passenger wandered onto the aircraft as if nothing was wrong. It was the same guy. This is why setting fire to people should be legal under certain circumstances.

Anyway, over time I became pretty stubborn, and I usually insisted on closing the entry door, finding the offender's bags (a legal requirement), removing them from the hold and then putting them behind the main wheels so that they got crushed flatter than a sheet of cling wrap when we pushed back from the terminal. (Okay, okay, we didn't really crush the bags, but we would have delighted in thinking about it.)

Sometimes while the ground crew were searching for the bags, the passenger would turn up and the ground staff would call us to ask if they could open the door and let them on. My answer was usually no unless their name was Jenna Jameson or Kate Hudson. While this bag search was going on, it was my job to get on the public address and inform the passengers as to what was going on.

'Good morning, ladies and gentlemen, and any Americans we might have on board. We're all ready to go, but some vacuous moron has managed to lose themselves in the terminal, probably in the bar. We're currently searching for their bag, and when we find it, it will be removed and crushed thinner than an IHOP pancake. We'll be underway shortly. Thank you for your patience.'

YOUR CABIN CREW

Note: all genders of cabin crew are now known as flight attendants. Hosties, stewardesses and the like have been consigned to the PC bin.

Before we start talking about cabin crew, I need to come clean and admit to having married a flight attendant. I often use this as an example of just how dangerous aviation actually is. At the time it was all quite romantic. We met over breakfast with the crew at the Hilton Mainz in Germany. The hotel sits on the side of the Rhine River. It was the middle of winter and the ground outside was covered in snow.

Manola wasn't even on my crew. After breakfast, a big group from different crews decided to head into Frankfurt to have a look around the city. Somehow, Manola and I managed to miss the train that the other crew got on. While waiting for the next train in −10°C, I noticed Manola was shivering so I put my arms around her to keep her warm.

A day or so later, I got a roster change to fly back via Bombay (Mumbai) with a three-day layover there, and guess who was on my new cabin crew? Fate works in mysterious ways. We'll talk about flying relationships a bit later in the book.

Even though you might think cabin crew are just there to serve you lunch and a tiny bottle of wine that would be better used as paint stripper, they are primarily there to save your arse in an emergency. They are generally a fine bunch of hard-working professionals and deserve to be treated that way. They work much harder than you think and must stay patient and pleasant even when they feel as uncomfortable and tired as you do. I have a lot of respect for cabin crew, even more so for the older ones from around my generation. I can't defend that last statement, but perhaps it's because the older ones are more patient with me. It's more likely just a gross generalisation made by a grumpy old captain, and I apologise if it offends. Actually, I don't.

When cabin crew first join an airline company, they undergo a few months of intense training with a very high pass/fail standard. They concentrate a lot on safety and need to know where every piece of emergency equipment is on every aircraft they will operate on. Unlike pilots, cabin crew can fly on many different aircraft types. As well as safety, cabin crew need to learn how to deal with difficult passengers, handle medical emergencies and serve meals. It's not easy asking, 'chicken or beef?' 300 times in a row without tripping over your words or losing enthusiasm.

There's a saying that everyone becomes instant royalty when they board an aircraft – in their minds, that is. But, even if you *are* royalty, or possibly first and business class, be humble and polite, as that will get you the best service. Dressing well also helps. Why? If the flight is oversold in economy and they need to upgrade someone, they certainly won't upgrade you if you are wearing thongs and a t-shirt. Dress well and you might also notice slightly better treatment; respectable dress earns respect, after all. By dressing well, you'll immediately stand out, and the crew will appreciate your extra effort. That might sound silly, but it's backed by science. Research shows that people make positive assumptions about individuals based on how they dress.

Your cabin crew have also heard every reason for a person to be upgraded to business or first class. They have listened to every pickup line and every dad joke (the dad jokes mostly from me). But still, they will indulge you by smiling. That's impressive.

Probably the most important job cabin crew do is to get you off the aircraft in a hurry when death is imminent. This is called an evacuation. The first recorded aviation-related emergency evacuation was in 1793, when Jean-Pierre Blanchard used a parachute to escape from his hot air balloon after it ruptured. The poor bastard had to do it all under his own steam too. He didn't have the luxury of a cabin crew on that flight.

These days, you will see evacuations on the TV news from time to time. A wheel collapses or something else fails and all of a sudden, parts of the aircraft are on fire. Then the next thing you will hopefully see is doors opening and slides inflating, followed by people jumping down the slides and running like hell, in circles, or in any direction that seems like a good idea at the time.

The most likely reason for an evacuation would be a fire inside or outside the aircraft. If there is any doubt at all about how safe it is to remain on the aircraft, the crew will order an evacuation. As uncommon as this is, it pays to engage your brain if it ever occurs while you are on board. More on this later in the emergency landings chapter.

So, your cabin crew are very important. I feel for cabin crew these days because they work hard rosters, and it must be challenging to have any semblance of a normal life. In many cases, the job is their life, though, and I suppose, if you enjoy it, then why not.

For now, it's time to move on, but throughout the journey of this flight, you will see the vital role the cabin crew play.

YOUR PILOTS

'*W*hen I had 50 hours of flying experience, I thought I was cool. When I got 200 hours and a commercial licence, I was, in modern-day terms, 'The Shit'. When I got my first 1,000 hours, I realised how little I knew at 200 hours, but at least now, I knew it all. Now that I have 23,000 hours, I realise how much I still have to learn about aviation.' (All old pilots)*

Having been a pilot most of my life and having met thousands of other pilots, it's interesting to hear what inspires young people to pursue a career as a pilot. In general, though, just about every pilot I know always wanted to be one from a young age. It's the sort of career that needs a great deal of focused effort to be successful, and it would be difficult to do it justice if you attacked it in a half-hearted way.

As for me, I got inspired by listening to the moon landing on a transistor radio when I was ten years old. I decided then and there that I would be an astronaut. It took a few years after that to work out that a New Zealander was never going to be an astronaut. The next best thing was pilot.

Airline pilots come through a few different pathways. Some join the services and get to fly military aircraft. Others go through training academies and find flying jobs as instructors or as commercial pilots working in a wide range of jobs worldwide. My own journey started as a navigator in the Royal New Zealand Air Force. Then I flew light aircraft in the mountains of Papua New Guinea before joining Qantas in 1984.

Some airlines run their own training schools. In my experience, no way is better than the other for the standard of pilot that is produced. In the longer term the airlines mould their new pilots to suit their system anyway.

And so, we get to the discussion about what makes a good pilot. The first thing to consider is that aviation is a dangerous game if not taken seriously.

The consequences of a crash can be extreme. A pilot's job is to anticipate risky events before they arise and recognise them if they do occur. They then need to mitigate the risk until it isn't a threat. The ability to do this is learned from training and experience. The more you have seen, the better you are at knowing what's coming. This is why experience counts, and why airlines and insurance companies have minimum hours requirements for young pilots to be employed. Unfortunately, accountants and airline management don't seem to understand or value this basic concept.

I've had times as a captain where I've taken some action in advance, which has been precisely what the doctor ordered. The younger pilots have asked, 'How did you know that was coming?' I would say, 'A gut feeling because I've seen it before.' An example of this might be where the weather forecast is good, but you have a gut feeling there's going to be fog. The weatherman can't see it, but your knowledge of a place tells you otherwise. Other times I've taken actions that later seemed a waste of time because nothing happened, but this is not wasted energy at all. There's a saying in the pilot world: 'If you are prepared for something, it will never happen, but if you aren't prepared, the likelihood of it occurring goes way up.' This is Murphy's Law.

Always being aware of the 'big picture' is critical. Pilots who focus only on the moment are not doing their jobs. On a typical flight, the big picture includes everything that could affect the flight within a large area of the Earth's surface in the shape of a racetrack. The racetrack includes the departure point, the final destination and any emergency alternate destinations. As the flight progresses, the racetrack stays the same shape in front of you but gets smaller behind you, as if the back end of it was being towed by the aircraft, because things happening way behind don't matter anymore.

In the cruise, it's the pilot's job always to know what they would do under different scenarios and always to have a plan in their head of where the nearest safe airport is. At the beginning of the flight, you need to be aware of the weather at any airport you may have to use, plus the en-route weather.

You also need to be mindful of any other technical or non-technical issue which could affect the flight. On international flights, you need to be conversant with the geopolitical situation in the countries that you are overflying. Yes, reading the news is part of the job. Knowing where tensions are rising means it won't be a surprise if airspace over a certain country or area is suddenly closed. You must always have a plan.

Nothing should ever be a surprise, except perhaps an offer of prawns and gateau from the first-class section. A good pilot will always ask themselves, 'What happens if I have an engine failure right now? What about a fire on board? What is the weather trend at the destination? What are the chances of some nutjob firing some missiles at his neighbour as a gesture of goodwill? How is the fuel remaining looking compared to the plan, and I wonder if there are any leftover prawns?' Effective crews discuss these scenarios as they go. This benefits the senior pilots as revision and the junior pilots as learnings.

Yes, good pilots definitely like to keep one eye on the big picture, and in the case of an Air New Zealand captain mate of mine, that's precisely what he did – because he only had one eye. The other was ruined by a popping champagne cork halfway through his career. Forty years ago, pilots needed uncorrected 20/20 vision to get a start in aviation. These days you could turn up with your head missing, and this would be okay as long as you wore your hat. Company uniform rules are very strict.

Of course, the big picture also includes the economic operation of the aircraft. With a fuel load of 320,000 litres, 26 crew and nearly 600 passengers, an A380 on a long international flight can have a $1 million budget every time it gets airborne. Decisions by the pilots can cost or save tens of thousands of dollars on each flight. A good captain will easily save his wage each year in operating costs. Sadly, the only reward for this is personal pride. Bonuses only go to those who least deserve them.

Pilots are actually, first and foremost, energy managers. How so? Energy management is the same for both a tiny aircraft and the behemoth Airbus A380, but the scale is different. Let's talk about the A380 because the figures are more impressive. All loaded up and ready to go, you have 320,000 litres of fuel, which is chemical potential energy. All you have to do is pump it into the engine, set fire to it and the engine produces thrust. This thrust moves the aircraft, producing kinetic energy. So off you go down the runway hoovering fuel at close to 10 litres per second to create more kinetic energy and some residual noise and heat energy.

Here's where things get a bit complicated, so I'll type slowly so you can keep up. Given that the formula for kinetic energy is half times the mass times the velocity squared, velocity (or speed) is the most significant contributor to kinetic energy. So, by the time you reach the cruising level, and you're doing 950 kph or 500 kt, you have a super tanker load of kinetic energy and a little less fuel. Actually, a 500,000-tonne super tanker travelling at 15 kt has a similar kinetic energy to an A380 travelling at 500 kt.

Now that the A380 is in the cruise, it has a bit less fuel, the super tanker of kinetic energy and something else called gravitational potential energy, which the aircraft has gained from climbing to high altitude. Gravitational potential energy plays out when the aircraft starts its descent to the destination and can basically glide at idle power from over 200 km out.

Descent is where energy management really comes into play. A straight-in descent with no interference from air traffic control still takes a lot of thought, because the aircraft must be organised to arrive at the final approach at the correct approach speed and the right height, and configured with flaps out and wheels down for landing. On the final approach, the speed is reduced to touch-down speed, a landing is carried out, and then it's the job of the brakes to reduce the aircraft's speed down to taxi speed. This creates a lot of heat energy in the brakes. And of course, the real trick is to never run out of fuel before landing.

But of course, the arrival is never that simple. Air traffic control can challenge the energy management of even the best pilots with vectors, speed restrictions and holding patterns. Good pilots see this as a challenge. Others get stressed out about it. Some pilots can't do it very well, but they don't know it.

While working for Japan Airlines, I used to occasionally fly with an American first officer who had no idea just how incompetent he was. On one descent into Honolulu from Japan, he started to become very high on the arrival profile. Once you get too far above the target profile, there is nothing you can do about it except do a circle or a large dog leg, both of which are very inefficient. So, I mentioned that I thought he was getting high. He scoffed at me and did nothing about it. I decided to let it go and teach him a lesson. Eventually we got so high that we had to ask for a circling manoeuvre to lose height. The problem was that there were other aircraft coming in behind us, so by the time we got back into the approach sequence we had flown for an extra 15 minutes.

I doubt he actually learned anything as he really was exceptionally self-absorbed. We had cost the company a lot of money. From a captain's perspective, that's not acceptable. At least I learned something: take over before it's too late.

Now, having discussed the two main ingredients of the recipe for a pilot, we need to talk about the herbs and spices that should be added.

Pilots must be able to prioritise tasks effectively and push less critical tasks to the back of the queue. Knowing what is unimportant and putting it aside for later allows you to focus better on what is important.

In addition, they must use the automation capabilities of the aircraft to take a load off their minds and allow brain space for effective decision-making. Using the autopilot to fly the aircraft means most of your brain space is available for managing. It's a highly effective way to operate while under stress.

Pilots should have good hand-eye coordination. But, having said that, one helicopter pilot I knew couldn't catch a ball.

They must have the self-discipline to turn up to work in good shape and with a clear head, no matter what is going on in their personal life. Turning up hungover or with emotional life stress can make a crew member less effective.

They must have good mental maths skills to crosscheck what the computers are telling them without grabbing a calculator. Computers are not infallible and are highly susceptible to incorrect inputs. Good pilots always check computer-driven indicators for logic in the real world.

They must be able to concentrate for long periods and stay focused even when exhausted. Quite often, pilots will be up all night and have to land in bad weather at a strange airport early in the morning. It's no good being half asleep. The ability to perform to a high standard in these conditions makes you valuable as a pilot.

For effective crew management, pilots must try not to be a dickhead. I haven't personally tried this yet, but I've heard it helps a lot.

They must be good leaders and managers, which is sometimes a tall order.

Pilot personality types have been the subject of many studies. A common misconception is that we are all 'type A' personalities. That's not correct. A 'type A' on its own may be dangerous. It's more a combination of A and B that's required. Studies have concluded that a typical pilot would be low on the characteristics of *anxiety, vulnerability, anger, impulsiveness* and *depression*.

But they would have an excess of *conscientiousness, deliberation, achievement, competence* and *dutifulness*. Pilots also tend to be *trusting* and *straightforward* and have *high assertiveness*.

But don't overthink all of the above. Overthinking things is not a good characteristic of a pilot either. And neither is dithering, which is one of my favourite words. While continually assessing risk, you must be able to 'get on with it'. I once told my wife, 'If you were a pilot, you would never get airborne. You would continually find reasons not to, and you'd constantly be distracted by unimportant stuff.' From memory, that was when I learned to cook and wash my own clothes.

And so now you are probably starting to get the impression that pilots are made of stern stuff and should be paid more. Well, yes, but in all honesty, they come from all walks of life and all 'edges of the spectrum', including outside the edge. There is no doubt that a particular skill set is required to be successful, and in my time in aviation, I felt lucky to work with such an exciting and talented group of people.

But like any profession, a small percentage didn't quite fit the mould. Everyone knew who they were too, and many had nicknames. Let's run through a few of the nicknames:

The Poison Dwarf (Personality and height related)
The Three Dwarfs (Sleepy, Dopey and Grumpy . . . and this
    was just one guy)
Two Dads (A double-barrel surname)
Duck man (He had a pet duck that was eaten by the
    next-door neighbours while he was away on a trip)
The Prince of Darkness (Former head of training, much
    feared by the weak and incompetent)

Captain Crime (Because crime doesn't pay . . . after dinner)
Beaujolais (Young and immature and leaves a nasty taste
    in your mouth . . . he's got better with age, as you would
    expect)
Wang (Surname: King)
Duress (What he put his crew under)
Delay (Name and behaviour related, in a good way)
Tony Barber (Always asking questions, like the *Sale of the
    Century* host Tony Barber)
Dirtbag (That's Mr Dirtbag to you)
SLAP (Short Little Angry Prick)

The odd pilot was disliked by everyone, and there was a collective sigh of relief every time one of the exceptional cases retired. However, I was always quick to point out that there would soon be a new pain in the arse coming through the ranks. All of this does not reflect these outliers' ability to do the job. They had to front up for checks like everyone else. As long as they were above the minimum standard, which is still relatively high, everything was okay.

And then there were the crazy but likeable ones. One of our captains was a slightly crazy but fun guy who, by all reports, flew well enough. I made a comment to some pilot on a flight that this captain was 'voted most likely by his peers to be seen on TV sometime in the future, naked in a clock tower with a high-powered rifle'. A couple of weeks later, the captain in question approached me in the crew room and said, 'Ah. Jeremy. I heard what you said about me, and I love it.' He shook my hand and walked off smiling.

In another more recent development in 2023, a former Qantas pilot was fined $4,500 and given a year's good behaviour bond for pretending to be a lawyer. After flying planes for 20 years, he took early retirement in 2020 seeking to 'move in a new direction'. And in 2021, a Jetstar pilot was charged with murdering two campers in Victoria. So as a group, pilots are just the same as any other group in the varied personalities involved.

Grumpiness goes with the territory of being a senior pilot, especially as age goes up and testosterone/oestrogen levels go down. Perhaps this is related to weariness and cumulative jetlag as well. Airlines usually contribute to pilot grumpiness, and so it's ideal, but not mandatory, that pilots start with a low level of grumpiness so as to not reach too high a level later.

An example of how airlines contribute to pilot grumpiness is with uniforms. Once every few decades, someone in the airline decides to make a name for themselves by proposing a new uniform to 'refresh' the product. This happened a few years back with Qantas. In the old days, the pilots would have been consulted as to what works best with a uniform, considering the long nights sitting up and trying to stay comfortable. And so, in keeping with tradition, the pilots were consulted, and then ignored.

Next, a fancy designer got paid a truckload of money to design the new uniform. The old uniform was black trousers, white shirt, plain black tie and gold epaulettes. The new undoubtedly mega-bucks design was black trousers, white shirt, plain black tie and gold epaulettes. But to improve the look, they added cuffed sleeves and random buttons to the shirt that made it impossible to iron. They also added a pocket that you couldn't stick a pen into, as it wasn't actually a pocket. It was a decoration that looked like a pocket. The new hat was so retro that it looked like a hat an ice-cream salesman would wear, thus making walking through the terminal during school holidays a right pain. Feedback from the pilots who trialled this new design was all very negative, so the company decided to proceed anyway, to piss them all off. This tactic worked.

At this point, the design was sent to the accountants for cost cutting. This involved getting most of it made from synthetics so that it would be both uncomfortable and likely to melt into your skin in a fire situation. And the gold epaulettes were halved in width to save gold. They also decided to remove the pocket from the rear of the trousers but make it so that it looked like there was a pocket there. The uniform was to be made in a sweatshop in Bangladesh, rather than providing work for Australians. The one positive outcome from all this was that being made from synthetics, pilots soon discovered you could tip a hat on its back, fill it with ice and cool your beer down, without it leaking.

If I had a dollar for every time someone, on hearing I was a pilot, asked if I had 'a girl in every port', I'd be very wealthy. The truth is, I doubt that pilots are any different to other professions when it comes to this sort of thing. Temptation is everywhere. The only difference for pilots is the opportunity when far from home, although the rumour mill is pretty efficient. It's said that a pilot's partner knows when they're going to play around even before it's crossed the pilot's mind to do it.

That said, it was not unusual to hear stories of pilots getting themselves in the poo. Some of the stories are standard fare, but others are legendary. There have been a couple of pilots I knew who were having affairs with flight attendants, so they'd pack a bag, get dressed in uniform, kiss the wife goodbye and go to work. Except they weren't going to work. They were catching a cab around to stay with the girlfriend.

Both pilots I knew who were doing this eventually came undone when crew scheduling called home looking for them. The wife would initially suggest that the schedulers were fools because they should know that the pilot was at work. The schedulers would duly inform the wives that the pilots were on days off, and the shit would hit the fan. The advent of mobile phones has added another safety layer for pilots playing this game.

Another of our famous pilots was having an affair with a flight attendant, who decided to come around and inform the pilot's wife that she was in love with him, and that the wife should move on. Luckily for him, when the poor girl turned up, he was out the front watering his garden. He turned his hose on her, and she jumped back in her car and left without the wife finding out.

Some 'incurable romantic' pilots are on their fourth and fifth marriages due to bad judgement. For a group that is supposed to be good at recognising risk, this is not encouraging. One particular captain tried to fast-track this multiple marriage deal by having two families at the same time. One was in Sydney, the other in London. You have to admire the administration skills needed to pull this off. He wasn't well liked by fellow pilots, so there was a lot of talk (wishful thinking) of him being locked up for bigamy. I suspect the truth is that he wasn't actually married to the London girl, as there were no reports of his incarceration. You can be sure that if there were photos available of him dressed in an orange prison jumpsuit and holding hands with a big dude called Bubba, they would have been circulated. The astounding thing, if you knew the man, was that there were two people in the world who wanted to sleep with him. I guess even hobbits need love.

Many pilots are heavy drinkers. A few are undoubtedly functioning alcoholics. There would be functioning alcoholics in all
high-pressure professions, so it's not just pilots. But they front
up and do their job, so good on them. Unfortunately, the job is
unhealthy because pilots often find themselves away from home
with nothing to do but go out for drinks with the crew.

Pilot managers are a good source of interesting stories about
the pilots they have to 'mother'. One manager friend of mine has
told me many stories, confidentially of course, about looking after
the pilots. Our managers are the ones who mediate between us
and the 'evil empire' of the human resources department, sometimes losing the battle. They also look after our general wellbeing
to make life easier for us so we can concentrate on the task at
hand: safely flying the aircraft without worry or distractions.
As a result, they get to hear our deepest, darkest and, at times,
dirtiest secrets.

On one occasion, one of the younger pilots approached their
manager and requested a roster favour. This was not for a day off
for a family occasion, but instead he asked to rearrange months'
worth of roster so he could avoid flying with a flight attendant
he'd had a dalliance with. His wife had found out and insisted
on checking his roster four days prior to flying. If she was to
see that particular flight attendant's name on the roster, she
would go ballistic. Rosters were often changed without anyone
ever knowing.

Then there were the times when disgruntled ex-girlfriends of
pilots would call and make allegations of drug and alcohol abuse
by the pilots, knowing management couldn't ignore it. On some of
these occasions, investigations would have to take place, meaning
weeks off work for the pilot. Of course, these stories are about

male pilots, but with the increasing number of female pilots joining the airlines, there will undoubtedly be some good stories about the other side before long.

The first time I met my last pilot manager, I had commuted in from Auckland to Sydney and needed to iron my uniform shirt before starting my duty. There was an ironing set-up in the crew room. I walked into my manager's office and said, 'Hello. Who is on ironing duties today?' I teasingly asked her to do her job and iron my uniform. She politely told me to fuck off and do it myself, and we've been friends ever since. She later won me over fully by buying me a sprinkle cupcake for my birthday. She knew we pilots were simple. The company spent millions fighting us through the unions when all we ever wanted was banter, kindness and cupcakes. Incidentally, I would never let anyone else iron my shirts. It's one of two psycho things I brought with me from the air force: never be late and no one can iron a shirt like I can.

We'll speak much more about pilots later. Right now, though, we need to keep moving or there'll be a delay, and nobody wants that.

YOUR AIRCRAFT

Now that you have been welcomed aboard by your wonderful cabin crew, it's time to strap in between random strangers. But what is it exactly you are strapping into?

Depending on where you are going and which airline you are flying with, you will most likely be on an Airbus or a Boeing. Airbus aircraft are made in the European Union and Boeing are made in the USA. The most common aircraft are as follows:

Airbus A320 series. Small, 2 engine, single aisle, carrying 220 passengers.

Airbus A330 series. Wide body, 2 engine, twin aisle, carrying 280 passengers.

Airbus A350 series. Wide body, 2 engine, twin aisle, carrying 400 passengers.

Airbus A380 series. Wide body, 4 engine, twin aisle, double deck, carrying 525 passengers.

Boeing 737 series. Small, 2 engine, single aisle, carrying 180 passengers.

Boeing 747 series. Wide body, 4 engine, twin aisle, double
deck, carrying 470 passengers.

Boeing 777 series. Wide body, 2 engine, twin aisle, carrying
395 passengers.

Boeing 787 series. Wide body, 2 engine, twin aisle, carrying
330 passengers.

Keep in mind that the passenger numbers are approximate and
can be increased in high-density configurations. For example, the
Airbus A380 can carry up to 853 passengers in an all-economy
configuration. Of all the above aircraft, the A350, A380, B777
and B787 are the longer range aircraft. The ultra-long range
version of the A350 is capable of flying almost 18,000 km.

A modern jet aircraft is some of the most advanced technology you'll ever be exposed to. Every new generation gets better
than the last. Take the Airbus A350 as an example: 52 per cent
of it, including wing surfaces and spars, is made of carbon fibre
reinforced polymer, which is lighter, stronger and more flexible.
Another 14 per cent of the A350 is made of titanium. This combination makes the A350 14 per cent lighter in empty weight per
seat than the older B777.

All this makes the accountants and shareholders very happy
as new aircraft use less fuel. Airlines will make announcements
about how this reduces emissions, but, in reality, they are rubbing
their hands together because it reduces their costs. All significant
decisions are made on a cost basis, and all major announcements
are made on a what-people-want-to-hear basis. It's just how the
airlines work.

Engines for these modern airliners are big business. A new
GE9X engine for a B777X is quoted at around US$42 million.

Keen observers of aviation will have noticed that they need two each, typically one on each wing. The cost is indicative of the expensive materials and the development of these high-performance engines. They are impressive power plants. The GE9X can develop 50 t of thrust, which means they can lift a weight of 50 t if pointed straight up. This also means they can lift nearly six times their own weight. They need to be this powerful because if an engine fails at a critical time on the runway at maximum take-off weight, the remaining engine must still be able to get the aircraft safely airborne.

Let me explain how a modern high-bypass aircraft turbofan jet engine works in simple terms. Imagine it as a combination of a jet engine and a massive fan. At the front of the engine, there's a huge fan, enclosed in a casing called the cowling. This fan is responsible for sucking in a tremendous amount of air from the atmosphere.

Now, here's where the 'high bypass' part comes in. The majority of the air that's drawn in by the fan doesn't go through the engine core, but instead bypasses it. This means it flows around the outside of the engine through a bypass duct. On a GE90, for example, 90 per cent of the air bypasses the internal combustion area.

As the bypass air flows around the engine, it is accelerated by the spinning fan blades. The bypass air is pushed out the back of the engine and, thanks to Newton's third law of motion, the 'equal and opposite reaction' pushes the aircraft forward. This accelerated bypass air contributes to a significant portion of the total thrust generated by the engine.

But what about the air that does go through the engine core? Well, this air is compressed, mixed with fuel and ignited in the combustion chamber, just like in a regular jet engine. The burning fuel creates hot gases, which then expand and rush out of the engine core, producing thrust. On the way out, it passes over turbine blades, which drive the main fan and the compressor blades.

So, in a high-bypass turbofan engine, you have two sources of thrust: the bypass air, accelerated by the fan, and the hot gases from the engine core. The combination of these two thrust sources makes high-bypass turbofan engines incredibly efficient and powerful.

The benefit of having a large amount of air bypass the engine core is that it increases fuel efficiency and has a quieter engine operation. It also helps to improve the engine's overall performance and reduce emissions.

It's also worth noting that the speed of the main fan on a GE90 reaches up to 2,355 rpm, but the internal compressor rotates at a much higher speed of up to 9,332 rpm. That's 155 revolutions per second, which explains why such high quality and highly precise components are required when building these engines.

It's inevitable that some of you petrolheads will be wondering how much horsepower we are talking about here. The bad news is that it's not directly compatible, so there is no easy conversion. It depends on many things, and the answer is somewhere between 55,000 and 90,000 hp. If that's not good enough for you, you can work it out using the formula below. The rest of you, feel free to skip ahead or get a cup of tea.

1. *Power of a static engine (P) can be computed from the kinetic energy imparted to the exhaust gases. This would be ½ × (mass flow rate) × Ve² (Ve is exhaust velocity).*

This will give you power in watts, which you can convert to horsepower using Google.

2. *The mass flow rate, assuming exhaust gases are at atmospheric pressure upon exiting the nozzle, would be air density × (exhaust nozzle area) × Ve.*

3. *Bring 1 and 2 together: $P = (\frac{1}{2}) \times$ (air density) × (exhaust nozzle area) × Ve^3*

4. *Density units must be kg per cubic meter, nozzle area must be in square metres, and Ve must be in metres per second.*

5. *Since exhaust gases are usually hot, air density will be lower than normal. You can estimate the density with the ideal gas law for dry air: (air density) = (air pressure) / ($Rd \times (273 + Tc)$), where air pressure is in Pascals or N/m^2 (100,000 at sea level), Rd is a constant = 287 J/kg/K, and Tc is the exhaust gas temperature in °C.*

6. *So, add 3 and 5 together to get your estimated power in watts, then convert to horsepower, and you're good to go.*

How did you go? Welcome back to all you others. To finish this initial briefing on what you are strapped onto, we need to talk about fuel. This is mainly because I worked as a refueller, and if you are good with a hammer, everything presents as a nail.

Jet fuel is kerosene. There are several reasons why kerosene is used in aviation. The first is that it has a lower freezing point. Jet aircraft operate for long periods at high altitudes where temperatures get as low as around −56°C. Jet A1 has a freezing point of −47°C, but, generally, because of the warming of the aircraft wing surface from surface friction, the fuel doesn't get close to that cold in the tanks. If it does – which happens sometimes – then action needs to be taken.

This happened to me a few times over the years on the long polar routes where temperatures are colder at lower altitudes. It's made worse if you depart from a cold place as the fuel is cold even before it gets on the aircraft. A few more hours of cold soak from the outside air temperatures, and you start to have a problem. The fix is to descend to an altitude with a higher air temperature, or speed up so that friction from the airflow heats the wings a bit more. Sometimes you need to do both.

The second reason for using kerosene is that kerosene has a higher 'flash point' than gasoline. That's a good thing. You can trust me on this or do your own research. Third, kerosene has a lower viscosity than gasoline, so it's less likely to clog things. Lastly, it is much cheaper than gasoline, so it gets three cheers from the accountants.

Each new generation of aircraft has better safety features and more redundancies in its systems. Lower in-flight cabin altitudes (which we'll talk more about later), bigger windows and cleaner air enhance passenger comfort. In addition, aircraft accumulate dirt as they age, so newer aircraft logically have less lying around. This means an airline with a younger fleet is the better bet for you.

When a company like Boeing or Airbus decides to build a new aircraft type to sell to customers, they must spend years designing, building, testing and certifying the aircraft before it can be used to carry passengers. Certification is the process by which manufacturers prove to regulators that the new aircraft is fit for purpose. This is generally all about safety. The process can easily take many years. For example, the A380 took nearly five years and involved 2,600 hours of flight testing and five test aircraft.

In preparation for clearance to perform the first flight, a new aircraft undergoes structural static tests that include a re-creation of the aerodynamic loads that the aircraft will experience during its lifetime. The tests include maximum bending of the wings, fuselage pressure tests and fatigue tests. They also include the simulation of flight cycles encountered from take-off to landing.

The next steps are certification tests to cover limit loads and ultimate loads in various situations, such as maximum wing deflection. Fatigue testing examines how the aircraft's structure responds to loads in conditions that occur in repeated flight cycles from take-off, climb-out and cruise to the descent and landing.

An aircraft's flight test campaign is designed to assess general handling qualities, operational performance, airfield noise levels, systems operation in normal flight modes, failure scenarios, and severe conditions such as extremely cold or hot weather and strong winds and rain.

Aircraft certification is a big deal. However, it can lead to disaster if it's not done well, as we saw with the Boeing 737 MAX: Lion Air Flight 610 crashed in October 2018 in Indonesia, and Ethiopian Airlines Flight 302 crashed in March 2019 in Ethiopia. The two crashes killed 346 people in total and were ultimately blamed, in part, on a faulty automated flight control system due to rushing the new aircraft into service. Part of the problem was that pilots were not even briefed on how the new but faulty system worked and how to handle a malfunction if it did occur.

At the opposite end of the scale, we have old aircraft. When some aircraft reach a certain age, they are converted into freighters. The rules for operating these are somewhat more relaxed. I suppose this is because no passengers' lives are at stake, and the pilots are expendable. When I flew for Japan Airlines (JAL) in the 1990s, they had purchased their B747 freighters new, so this wasn't an issue. But the Japanese had made a deal with

Evergreen International to carry some of their freight, and, as a consequence, we JAL pilots had to fly sometimes as passengers on the flight deck of Evergreen freighters.

On the B747, we used a system where if something was broken, we'd hang a small flag off it to indicate that to the crew. Of course, you would only see a few of these at Qantas over a two-month period, and I never saw any at Japan Airlines, but the Evergreen flight deck had more flags than an Olympics opening ceremony. One ex-Evergreen captain I knew said that they had so many failures in flight that when they went into the simulator for their six-monthly checks, they practised normal operations, because they never got to see them otherwise.

The 747s have bolts that attach the engines to the wing structure. These bolts are called shear bolts because, under extreme circumstances, it's better for an engine to break off at the bolts and leave the wing structure intact. But when an Evergreen 747's engine fell off in turbulence and 'landed' in a supermarket car park in Anchorage, Alaska, while I was there in the 1990s, the bolts didn't shear. Instead, the front part of the wing broke off due to corrosion.

The moral of the story is to avoid old aircraft and any aircraft owned by nations that might not have the most robust maintenance requirements. So, if you are on such an aircraft reading this right now, order a stiff drink and good luck and, um, can I have your Super Bowl tickets?

So now, getting back to the A350 and its competitor, the B777X, which is of a similar standard, it raises the question: Boeing or Airbus? You've come to the right place to ask that, because I flew both the B747 and the Airbus A380. I'm going to take a stand and sit firmly on the fence here, and say both.

They use different flight control systems, but both companies are very good at making robust and efficient aircraft.

But enough about your aircraft for now, because we are nearly ready to go, and your pilot wants to have a word with you to settle your nerves and explain why you are late. They will introduce themselves and try to sound as godlike and confident as possible without being too cocky. There is an art to this. You can't sound too serious, or it will scare people. You can't be too flippant, or passengers will think you aren't serious enough. Conversational and confident are all you need. Waffling is not a good trait. Get to the point, then shut up. A little bit of humour is okay too.

Domestic pilots, who might fly four sectors daily, get very good at rattling off clichés at high speed during their announcements. My favourite overused cliché is, 'Sit back, relax and enjoy the excellent xxxxx service.' This reminds me of my mother, who used to serve me lambs' liver and say, 'You'll eat it, and what's more, you'll enjoy it.' But we are getting distracted again, and I know you are busy, so let's get back on track and talk about air traffic control.

AIR TRAFFIC CONTROL

The doors are now closed, so nothing is holding us up except air traffic control. We always have to ask them for an airways clearance, which is permission to fly on a particular route, and with any luck, it will be the same as our flight plan. Our company should have requested that. Sometimes it is different. Sometimes they change it while we taxi for take-off.

A change during taxi in the USA causes much waving of arms and pulling out of hair in the cockpit, because a trip across the continental USA involves at least triple digits of different airways and turning points. Airways are like roads in the sky between geographical points. The points, also known as waypoints, are often (but not always) over well-known towns or landmarks. If the track actually changes direction over a point, then it is a turning point.

A last-minute flight plan change can be quite stressful on many counts. Normally on New York and Los Angeles departures crossing the USA, we only have two pilots. One person always has to be in control of the aircraft. That leaves only one person to write down a new clearance and correctly read it back to air traffic control. Then the clearance has to be loaded into the computer, which means having your head down for a few minutes. Finally, the other pilot has to check it has been loaded properly. The logical thing to do would be to stop, park the brake and do it together, but this is normally not an option as air traffic control wants the queue to keep crawling towards the departure point.

So, when you're taxiing out for take-off in New York, and the controller says, 'Change of clearance. Are you ready to copy?' everyone's hair stands on end. When you gather your courage and quietly whimper, 'Go ahead,' the controller then reads the new clearance out at warp factor ten:

'Qantas 26 cleared to K, L, A, X via runway 13 right Getthe-fukoutofhere departure, Hangover transition, Alpha17 to Allun, Whisky 92 to Joice, Delta 16 to Issan, Mike 23 to Ashol, Charlie 10 to Sumfnplace, Tango 181 to Sumotherfnplace . . .' and so on for at least a minute.

These guys are in a hurry, and they think that the sign of a good controller is that they can speak at ten words per second, so a minute is an A4 page. I'm sure you get the idea.

Generally, air traffic control in most countries is well run and professional. In technologically advanced countries, its systems are modern and complex and designed for safety. It's their job to organise large numbers of aircraft to get from A to B without running into other aircraft. At any one time in the world, there are around 20,000 commercial, cargo and corporate flights airborne, so good air traffic control is a big deal.

Sometimes when it is busy, and there are a lot of aircraft in a piece of airspace, you'll hear the controller talking non-stop, and it's hard to get a word in. This must be pretty stressful for them, and once in a while, you will hear one of them start to procrastinate, trip over their words and get flustered. Then suddenly, an older, calmer voice takes over, and the defeated controller, who is worn out, hungover or under training, is dragged away for retraining or sold to a pet food company.

Many younger pilots take a while to learn that air traffic control is there to help if they need it. It's no good allowing them to put you under pressure by hurrying you up, as that leads to mistakes. And if you need something, ask for it.

I was flying into Rome in a 747 from somewhere in Asia in the early 2000s. We came up over Cairo, and to the south of Crete, we hit a headwind that was much stronger than forecast. By the time we started the descent into Rome, we had just over an hour of fuel left. I wasn't too concerned because the weather was nice, but when we reached a point 35 miles out, air traffic control made us enter a holding pattern. We went around the pattern once and then asked the controller how long the delay would be. He didn't know.

We went around the pattern again, and still nothing. I told the first officer to declare a fuel emergency, which he did. This should have gotten some action, but I imagine Rome air traffic control just shrugged, because nothing happened. I told the first officer, 'Tell them we are coming in,' which he did. Then I pointed the aircraft at the airport and headed in without a clearance.

Suddenly the Italians started cooperating, and we landed safely and taxied in. As a result of my actions, I expected to be locked up and made to eat pasta for the rest of my life, or at least to be invited for an interview and some paperwork, but nothing happened. Go figure.

TWO

TAKE-OFF

SHIT GETS REAL.

ENGINE START

After the crew receives an airways clearance, the pilot who is flying the jet on that leg will brief the other pilots on how they intend to carry out the departure. This will include contingencies for things going wrong. Next, you seek clearance from the ground controller to push back away from the gate and start the engines. They will give this if the tarmac is clear behind you, but you still can't do anything until all the checklists are complete.

Pilots use aircraft checklists to prevent complacency by following every step to ensure aircraft are correctly configured for every phase of flight. Missed steps and incorrectly remembered steps have contributed to many aircraft accidents or incidents over the years. There's a certain amount of self-discipline involved with using checklists, and cross-monitoring by other pilots adds to the integrity of the operation.

Checklists are being used more and more in industries where safety is paramount. But, interestingly, they only started being used by surgeons in recent times. A surgeon friend in New Zealand said they started using them there due to a recommendation from a pilot who was watching an operation. I find this quite astounding, and I'm relieved they are now using them. 'Right. That's the last stitch. Let's clean up and get out of here.' 'But sir, a pair of forceps is missing, and what's this spare liver doing here?'

Have you ever lain awake in the middle of the night going over the things you need to get done? Using checklists mostly stops this silliness. I use checklists at home. It's actually not that anal. It's about being organised. I use checklists for things I need to get done day to day and for packing for a trip so I don't forget anything. I have a checklist for camping holidays in a Word file that I can spit out on demand. I even have a flow chart to tell me if it's time to buy flowers, which is a relationship checklist. This flow chart is very clever, as it works through all the issues, and no matter which way you go, it always comes out at the 'buy flowers' conclusion.

Most of the checklists on modern aircraft are electronic and presented on screens in front of the pilots. In addition, many of the checklist items are connected to the systems they refer to. This makes it impossible to respond incorrectly to that checklist item, because the system won't allow the checklist item to be ticked off until the configuration is correct. This is a significant improvement on the old read-and-respond type of checklist.

If an emergency occurs at any time during a flight, depending on the type of emergency, some actions need to be carried out by memory, because time is critical. Pilots need to be able to spit out these actions while under stress, so they must know them well. Once the aircraft is under control, then it's back to the electronic checklist, and pilots need to take a lot of care to follow the checklist items in order, especially when one particular action results in more required actions popping up.

Sometimes just sitting on your hands helps when an emergency unfolds. This stops you from rushing in and making it worse. When I was a new pilot, an old engineer gave me some advice about not rushing and doing the wrong thing. (Don't read on if you are eating.) He said, 'When something goes wrong, don't touch anything. Stick your finger up your bum then put it in your mouth. When you get used to the taste you can do something.' Point taken, Bob.

Anyway, if we don't actually get these engines started, we won't be going anywhere, so let's stay focused. To start a large jet engine you need a source of compressed air. This is normally provided by the aircraft's APU, or auxiliary power unit, a small self-contained jet engine normally found in the tail of the aircraft. The compressed air from the APU is used to spin the compressors on the main jet engine. Once this gets to a point where enough air is being compressed and sent to the engine's combustion chamber, fuel and ignition are applied and a fire starts to burn in the combustion chamber. This allows the engine to keep accelerating to idle power, which completes the start. Once one main engine is started, compressed air bled from it is enough to start the other engine or engines. As the engines go through this start sequence, different types of noise can be heard. Before fuel is added, the run-up of the compressors is a higher-pitched noise, but once fuel is added you get that throaty rumble, which sounds so powerful.

TAXI

As a passenger sitting down the back, do you always heave a sigh of relief when the aircraft starts taxiing for take-off? I do, because it means all the systems are working, and the TAB has just changed the odds of your getting airborne. The longer you sit there with engines running and not moving, the more likely it gets that the captain will come onto the PA (public address) and tell you that, 'We have to go back to the bay to let the engineers check something and don't worry, we'll be on our way shortly, oh, and the first officer knows a shortcut so we should still arrive on time once we do get going again.'

Assuming all is well, and we do start moving, it's time for the cabin safety briefing. There is a legal requirement that this happens. Airlines pride themselves on outdoing other airlines with their wacky safety videos. If you are a frequent flyer and you've heard the same corny safety video 50 times, this becomes very tedious.

The problem with the safety demonstrations and videos is that you can scare the passengers if you are too serious. People who are already anxious about flying don't like being reminded of what could go wrong. Some of the more modern airlines like to add some fun into the safety demos to try and get your attention, and good on them; it works for me.

Cabin crews are rightly enthused about the use of video, because it means fewer of them have to stand at the front of the cabin on display. They know that passengers are checking them out for regrowth and errant nose hairs and giving them a mental score out of ten for looks. I wouldn't like that very much. However, the good news for the cabin crew is that a high percentage of passengers are too busy sleeping or fiddling with something to bother watching . . . unless the crew member is a ten.

If you're one of the ones not paying attention, all I can say is that in an emergency, it's every man for himself, jungle rules, don't hang back and good luck.

Also, at the start of the taxi out, the pilots will extend the wing flaps to the take-off setting. This has the effect of changing the aerodynamic shape of the wing to increase lift at lower speeds. If they didn't do this, the take-off distance required would be much longer and, in some cases, exceed the runway length available. In most aircraft, if you are sitting near the centre of the aircraft, you will be able to hear the whine of the flap drive motors spinning to extend the flaps.

In most places the taxi out should only take 10 to 15 minutes maximum. But in some places, at certain times of the day, it can take much longer. A good example of this is New York's John F Kennedy Airport. It's not unusual to be in a queue for take-off for an hour or more. Sometimes you might have 40 aircraft in front of you waiting to go. If air traffic control closes some important airspace or starts re-routing aircraft, the delays will only get worse. In icing conditions, it gets worse again.

I've often seen situations where aircraft in the queue have burned so much fuel sitting there that they have had to return to the terminal to get more gas. I've also seen aircraft sitting in queues with all of their engines shutdown. This takes some planning to make sure you are all up and running when it comes time to move. I've seen pilots fail at this as well, which causes more delays.

At airports like this, make sure you use the bathroom before you get on the plane. It's a long time from 'doors closed' to 'airborne and seatbelt signs turned off'. Assuming you are one of the clever ones who did that, you can relax and start a movie or a book, and before you know it, the excitement of take-off will be happening.

TAKE-OFF

For me, the take-off is the most exciting point of a flight. You've had this massive build-up of planning and preparing, and now, finally, you are about to launch. It's like the kick-off at an international rugby test match or the start of a Formula 1 race. In an A380, the feeling of 300,000 lb of thrust being released and the take-off speeds of nearly 300 kph never gets boring.

On take-off, the maximum available thrust is used only when absolutely required due to being at maximum take-off weight. At lighter weights, a lower thrust setting is calculated based on the runway length available and the wind and temperature at the time. The reason for this is that lower thrust settings reduce internal engine temperatures, thus increasing the engine's life.

Apart from increasing the engine's life, there's another financial penalty if you don't conserve thrust. In a good percentage of cases, engines are now effectively leased from the manufacturer rather than bought, so instead of the manufacturer selling engines, they are actually selling thrust. Part of a modern lease deal is that the leasing costs vary depending on the thrust usage of the engines.

This is monitored via satellite 24/7 at the home bases of the manufacturers. Effectively, using more thrust on heavy-weight take-offs more often means a bigger bill from the manufacturer.

This minimisation of thrust also applies in flight, where using a higher thrust to climb more quickly costs more as well. So, pilots are trained to use the minimum thrust when safe to do so. But this also comes under the 'personal pride' banner. Why would you intentionally do something costly and inefficient when you are part of a business?

There are a number of decision points in a take-off, and good crew coordination is required to keep the take-off safe. Generally, after a lower speed of, say, 80 or 100 kts is reached, specific aural warnings are inhibited until after take-off, and the pilots will only stop (reject) the take-off for significant problems such as engine fire or failure.

Rejected take-offs involve reducing the thrust to idle and applying brakes to bring the aircraft to a halt or a slow taxi speed. The faster you are going when the decision is made, the more stopping effort is required. Just before the final decision speed, a rejected take-off would involve maximum braking and full reverse thrust to bring the aircraft to a halt before the end of the runway. At heavy take-off weights, this maximum braking effort can result in brake fires and deflated tyres. For passengers, the rapid deceleration and noise can be very disconcerting.

I've had three rejected take-offs in my career. All have been at relatively low speed, which is lucky. My one claim to fame was that I had one at Sydney Airport in a 747 with Nicole Kidman and Keith Urban on board. The engineers fixed the problem in an hour and a half, and we loaded up minus Nicole and Keith, who'd decided to try again another day.

The next decision speed that is reached on take-off after 100 kt is known as V1. This is a speed which is calculated for each take-off based on aircraft weight, runway length and weather conditions. On a balanced V1, if an engine fails before that speed, the take-off must be rejected. To continue would probably mean not getting airborne before the end of the runway. If an engine fails after V1, the take-off must be continued, because to reject from there would mean not stopping before the end of the runway. Depending on what is off the end of the runway, this might result in a splash or a loud noise, neither of which would enhance the captain's future career prospects. There have been quite a few incidences of precisely this in the past.

After passing V1 and reaching 'rotate' speed, the pilot not flying calls 'rotate' and the pilot flying gently pulls back on the controls or joystick (Boeings have a control column; Airbuses have a joystick). This causes the nose of the aircraft to rise to 12 to 15 degrees above the horizon, sometimes more. It also has the effect of raising the relative angle of the wing to the oncoming airflow and increasing lift generation. The increase in lift allows the aircraft to get airborne, and then we are off to the races. Once airborne and stable, the wheels are retracted, accompanied by the cacophony of noises that goes with that. Usually a last 'thump' indicates the gear doors are closed and then the noise reduces considerably.

A heavy-weight take-off in an A380 takes about 45 seconds and burns over 10 litres of kerosene per second. Be thankful that you're sharing that cost with 500 other passengers. If you look out the window at the wingtips, you'll notice them begin to rise as the speed increases, and then at the lift-off point, when all of the weight goes onto the wings, the wingtips rise a lot more.

This varies on different aircraft, but it's incredible to watch how much the wings can flex. It's worth noting that during aircraft certification, they bend the wings up almost until the wingtips touch, and they still don't break.

As the main wheels lift off the ground, it becomes obvious to you that you are airborne, because you can't feel the rumbling of the wheels running along the runway, and your senses tell you that the nose of the aircraft has gone nearly vertical. The reality, is though, that the nose has only been raised to the 12 to 15 degrees above the horizon.

The aircraft senses that it's airborne as well. For this it uses air/ground switches. These switches can tell whether there is weight on the wheels or not. On aircraft there are many systems that only operate in the air and many that only operate on the ground. The air/ground switches ensure the right systems operate when needed. For example, reverse thrust should only be able to operate on the ground, so it's electronically locked out in the air, and pressurising an aircraft should only happen in the air, so the air/ground switches control that as well.

One of the things that pilots are very wary of on take-off and, indeed, in flight is running into birds. For some reason, birds like to hang around airports, especially airports near the sea or wetlands. No, I have no idea why, except that they are possibly aviation enthusiasts.

Flocks of birds have been known to cause engine failures and bring down large aircraft. Because of this, engines have to undergo bird-ingestion tests during certification. They must be able to take direct hits by birds of a specific size and still keep running for a certain amount of time after the impact. Given the nature of this testing, it is hard to find birds willing to volunteer for it, so they either use real birds that have already been prepared for the pot or gelatin substitutes.

The other thing birds can do is cause structural damage to other parts of the aircraft. Given the speeds involved and knowing the formula for kinetic energy, the forces involved are massive. A mate of mine had a largish bird hit the windscreen just after take-off once. He said it cracked the windscreen outer layer, frightened the daylights out of him, made a hell of a mess on the windscreen and in his underwear, and he also said that the bird looked just as unimpressed by it all.

Now I know that, because you are the smart and inquisitive type, you will already be thinking, are there window wipers on planes to clean off such a mess? Yes, there are, and surprisingly, given the speeds planes are doing, they work. Whether they are good enough to clear off the remains of a bird with bad judgement is another thing altogether. A good rain shower would probably do a better job.

A discussion on birds and aeroplanes would not be complete without mentioning the 'Miracle on the Hudson', where both engines of Captain Chesley 'Sully' Sullenberger's Airbus A320 were taken out by a flock of Canadian geese after taking off from New York's LaGuardia Airport. It was a very rare occurrence to lose both engines to birds, and Sully handled it very well given the minimal time he had to come up with a plan. There was also luck involved. Bad luck in hitting the birds, and good luck in having a long stretch of calm water to land on.

In the 2016 film *Sully*, directed by Clint Eastwood and starring Tom Hanks as Sully, some judicious stretching of the truth occurs. This is because, apart from the first bit where Sully lands safely on the water in the first ten minutes, there is no suspense and no villain. There's the heroic bit and then the investigation. An actual investigation by the USA's National Transportation Safety Board of this incident subsequently found that all crew had done the right thing.

So in the absence of a real villain or antagonist, the producers made them up. The majority of the movie covers a concerted attack on Sully and his first officer by the National Transportation Safety Board, and only at the end is Sully vindicated. This attack didn't actually happen, and as a pilot, it's very disappointing that the investigators were made to look so bad. As I watched the movie, I found myself extremely frustrated by the investigators to the point that it made me wary of all investigators. It was only later that I found out it was BS. That was quite a relief.

CLEAN-UP AND CLIMB

Once the wheels are up and tucked away, the pilots can choose to fly by hand for a while to keep their flying skills up to date, or they can engage one of the autopilots. Most modern aircraft have three autopilots to choose from for redundancy purposes. They should be named Eeny, Meeny and Miny as they are all exactly the same.

The aircraft usually has to follow a standard instrument departure (SID), which is a designated track designed to position the aircraft on one of the main airways leaving the departure point. In areas of high terrain, SIDs help aircraft avoid high ground. SIDs are also designed to avoid built-up areas so as to not wake up retired pilots having a nap.

If there is no need to reduce noise, the pilot (Moe) or autopilot will lower the nose at about 1,000 ft above the ground to allow the aircraft to accelerate and then bring the flaps up. If noise is a factor, this acceleration phase will be delayed until 3,000 ft above the ground. The acceleration phase is known as the 'clean-up' and involves making the aircraft aerodynamically 'clean' for high-speed flight. Even though having flaps extended increases lift at low speed, it also increases drag, so for high-speed, low-drag flights, everything needs to be stowed away out of the airflow.

At the completion of the clean-up, there is normally a checklist to make sure nothing has been missed, then the climb is monitored and directed by the pilot flying, while the support pilot talks to the different air traffic controllers on the radio. At this point, if you are busting for a pee or eyeing off the spare seats just up from you, the pilots will normally check that the sky is clear out the front and then turn off the fasten seatbelt sign. This is like the starter's gun going off. If you are slow out of the blocks, don't blame me. On airliners with spare seats, there are two types of people, the quick and the dead.

Part of the discussion with air traffic control is negotiating the initial cruising altitude. If there are a few aircraft heading in the same direction at the same time, this can get quite competitive. This is because aircraft at the same altitude on the same track have to be separated by many miles. Aircraft without this separation must fly at altitudes 2,000 ft apart.

Flying at the optimum cruising level for your current weight gives you the best range and lowest fuel burn overall. If you have to fly lower, then range drops and fuel burn goes up. Extended periods of flying below your optimum altitude can sometimes cause a diversion prior to your destination due to running low on fuel, so it's imperative not to be held down low if possible. Pilots who are really onto this will sense when another aircraft is about to ask for their desired altitude and get in first. When you do this, you can almost hear a collective 'D'oh' from the other flight deck.

Sometimes, if someone else gets in first with an altitude request, pilots will ask for the next higher altitude. You have to be careful with this, because being too high for your weight can be quite perilous in rough conditions. In addition, being too high too early burns more fuel, but not as much as being too low. And being positioned above your competition stops them from climbing when they want to later, so effectively you have won the battle of wits and the associated bragging rights. This jockeying for the best altitude needs to be sorted while you're still close to the departure airport and still showing on the controller's radar, as once you disappear off their radar, more strict separation rules need to be applied.

At some point on the climb, a transition altitude will be reached. This is a specific altitude at which aircraft change the way they operate their altimeters, meaning pilots switch from using the atmospheric pressure at the departure airport to a standard pressure reference. This altitude varies depending on what country you are flying in.

During take-off and initial climb, the pilots' altimeters measure the aircraft's altitude based on the actual atmospheric pressure at the airport or the runway. It helps them accurately know their height above the ground. However, as the aircraft continues to climb, the atmospheric pressure starts to vary due to different weather conditions and geographic locations. This can cause inconsistencies between aircraft altimeters and create confusion for air traffic control.

The standardised pressure setting is known as the 'standard atmosphere' or 'standard pressure'. It's a set pressure value of 1,013.2 hectopascals (hPa) (29.92 in Hg) that all aircraft use above the transition altitude, regardless of the actual atmospheric conditions. This helps maintain consistent altitude measurements and improves safety in air traffic control.

In the initial climb, because the air is more dense at low altitudes, the climb rate of the aircraft is relatively high. As the aircraft climbs and the air becomes less dense, the performance drops off until near cruising level, when climb progress can become glacial.

As the aircraft climbs from the ground to the initial cruising level at, say, 36,000 ft, the air pressure and density drop by more than 75 per cent and the temperature on an average day drops to −56°C. The composition of the air stays the same as far as percentage of oxygen, but there's a lot less of it overall. So, the time that you could be exposed to this before you passed out from hypoxia at 36,000 ft would be 15 to 30 seconds. This is why, if the oxygen masks ever drop down, you shouldn't dither. Put one on yourself first then take care of those around you.

So, the aircraft has to compensate for the change in outside conditions as you climb, or else you'd need to bring extra layers of clothing and spend the whole time sucking on a mask. It does this by pumping air into the cabin at the required temperature. This air is bled off the engines and conditioned by air-conditioning packs. The air also pressurises the cabin to keep the oxygen component of the air at an acceptable level as well.

In an older aircraft at 36,000 ft, the cabin air might reach 7,000 ft, which means it's the same as if you were sitting on a mountain at 7,000 ft. Older aircraft have to accept lower cabin differentials than newer aircraft. The latest aircraft, like the B787 and A350, because of construction techniques, can accept a higher cabin differential. This means the in-cabin altitude can be kept lower at, say, 4,000 ft, and hence the level of oxygen is also higher, among other things. This results in greater comfort for passengers and reduces the residual detrimental effects after the flight.

It would be impossible to completely seal the cabin air in like a balloon, and indeed this wouldn't work as the outside pressure varies so much. You may have seen the effect on a closed bottle of water on descent. It collapses in on itself. The aircraft cabin doing the same would not be ideal, so the cabin pressure is kept at the desired amount relative to outside pressure by regulating the release of it through outflow valves at the rear of the aircraft. These open and close to adjust the amount of outflow and hold the desired internal cabin pressure. In this way, the cabin air is renewed completely once every two or three minutes.

This refreshing of air every two or three minutes is a very good thing, particularly on the climb. Because the pressure in the cabin is slowly reducing from that at sea level, gases in your body expand and need to be let out. You probably know this as the need to fart, but it does have a scientific name. You can use this as a conversation starter with your fellow passengers: it's known as HAFE, or High-Altitude Flatus Expulsion.

You can't beat HAFE with willpower, so it's best to let it happen, knowing that you are not alone. The cabin crew learn quickly not to fight it, and because they are caring and sharing, they will just let it out as they walk through the cabin. This is known as 'crop-dusting' by the crew, and they don't even give it a second thought.

Because the air in the cabin is pressurised relative to the outside, there is effectively up to 8 pounds per square inch of pressure pushing outwards on the doors. But the doors are set in a frame and have to be opened inwards initially before opening outwards so, at altitude, there could be up to 24,000 pounds of pressure stopping the door from being opened. This is why you would be wasting your time if you decided to get a bit of fresh air and open a door.

Sadly, it is not uncommon for people to try. Whether suicidal, delusional or just plain ignorant, they won't achieve it, so rest easy on that one.

The other thing that happens in the climb, given that the pilots won't have eaten the crew snack, is that they start wondering when coffee and food will be offered. This wondering can happen even before the wheels are up, because it may now be some hours since they last ate. Usually, the offer happens fairly quickly, which is good, because grumpy and hungry pilots can't concentrate.

THREE

IN THE AIR

YOU WOULDN'T BELIEVE IT IF I TOLD YOU.

INITIAL CRUISE

After about half an hour, the aircraft will reach its initial cruising altitude. The indication of this is that the nose will drop a bit from around 5 degrees nose up to around 2 degrees nose up and the engine noise will reduce. Less thrust is required to cruise than to climb.

On the flight deck, if it's a long flight with multiple crew, a game of musical chairs will be taking place, whereby one or two crew members will take their seats in the cockpit for the first shift on duty while the others bugger off for a rest. The two crew on duty will make sure everything is set up for the next few hours and then begin monitoring and planning ahead.

The cabin crew will be getting ready to serve drinks and a meal, and you will be waiting with great anticipation to hear the soothing tones of the pilot's voice telling you everything you want to know, and lots of other details you don't give a shit about.

PAs from the pilots are an annoyance to some and a calming influence for others. This makes the pilot's job difficult, as it's impossible to keep everyone happy. The critical information people want to know is what time they are getting there, and that everything is proceeding as planned. For some passengers, even this is too much, as they had no idea what time they were due there in the first place. If there are time zone changes involved, these people are already in a state of confusion. Crossing the international dateline can make some people's brains freeze up like the 'Microsoft blue screen of death' on a computer.

Other people, fearful flyers in particular, want to know everything, especially how safe they are. Many pilots are not aware that about 40 per cent of flyers are nervous to some extent or another. The important thing for these people is that they are kept informed whenever something unusual happens, and particularly when turbulence starts. They need to be told that everything is okay and that they are in good hands, or they may try to open a door and take their chances travelling on foot or swimming.

Pilots need to put some thought into their PAs to avoid scaring the passengers. Pilots are aware of what the company wants them to say but are not well briefed about what not to say. It's a given that you should never ask, 'Are there any pilots on board? If so please make yourself known to a member of the cabin crew.' Referring to the lift running out or the wing disconnect button are also not encouraged. The list of what not to say is extensive, even if many pilots are unaware of it.

One danger of making a really long PA, apart from annoying the customers, is that it sometimes doesn't transmit to the cabin well. This can be because your microphone is clogged with cookies and coffee, or because the aircraft system volume is turned down too low. There are not many more annoying things for a pilot than reading *War and Peace* to the passengers, finishing the PA and immediately getting a call from the cabin crew boss.

'Captain, It's Raymond. Parts of your PA didn't come through overly well.'

'Which parts, Raymond?'

'All of it, Captain.'

One of the great things about PAs is how they can be screwed up to the extent that all aircraft in the vicinity get to hear them. This is because the pilot's microphone has a bunch of switches to select where the message will go. Sometimes a pilot will make a PA and mistakenly transmit it on the air traffic control frequency. They can also answer a call from the cabin crew on this frequency. I've heard pilots putting in a full order for breakfast over the radio to air traffic control. Once or twice, I may have even heard myself doing it. This usually results in all the pilots of the aircraft within 300 km adding on their requests for long blacks and croissants.

Although the embarrassment that comes from making this error is only temporary, it is fairly intense. When you are in your hotel room afterwards, rocking back and forth on a chair, it's important to be kind to yourself. Afterall, most pilots have done it or will do it in their careers.

THE INFLIGHT
ENTERTAINMENT SYSTEM

One of the reasons passengers hate PAs is because by the time they come, they interrupt your movie at a critical moment. When the pilot starts by saying, 'I'm sorry for interrupting your movie,' it's obvious they don't mean it, so that raises the hackles even further. These days you can normally start watching a movie the moment you reach your seat, but doing this means you will have to put up with multiple annoying interruptions, because all announcements for the departure freeze the movie. In addition, the safety video overrides your movie, which, if you think about it, isn't such a bad thing. You really should be paying attention.

Of course, all these interruptions weren't a problem in the years before inflight entertainment became available on aircraft. In those days, passengers would engage in conversation with fellow passengers, read books and do crosswords to ease the boredom. Sudoku wasn't a thing then. These days, even the inflight magazine doesn't get much of a run unless the inflight entertainment (IFE) is broken.

The first inflight movie, *Howdy Chicago*, was shown in 1921 on Aeromarine Airways on a scenic flight around Chicago. In 1936, the airship Hindenburg had a piano, lounge, dining room, smoking room and bar. This helped while away the hours during the two-and-a-half-day flight between Europe and America. Slowly over the years, offerings improved from movies projected on screens to personal video players and then, finally, to TV screens in the backs of seats.

The introduction of the IFE has not been without its issues. Safety was, and still is, a major concern. With the sometimes kilometres of wiring involved, voltage leaks and arcing are a problem. The IFE system was believed to have caused the crash of Swissair Flight 111 in 1998 off the coast of Newfoundland. A fire took over the cockpit which was thought to have started in the IFE. These days, to avoid any possible issues, the IFE must be isolated from the main systems of the aircraft. Most of the bugs have now been ironed out, and the worst that happens is that the whole system crashes and needs a reboot, or it crashes permanently and the advertisers in the inflight magazine finally get some exposure.

When modern IFEs are working, they are great. The moving map, movies, news and audio on demand are good passers of time. The problem is that if they break, passengers get grumpy, and many believe they should be compensated. A broken IFE also means more work for the cabin crew, as a bored passenger needs more looking after.

The cost of these systems is not insignificant. A full system for a large commercial aircraft could come to upwards of US$5 million. Large international airlines sometimes pay more than US$90,000 for a licence to show one movie over a period of a few months. These days, airlines can feature up to 100 movies at once, so the costs can really accumulate. In the USA, airlines pay a flat fee every time a passenger watches a movie. Some airlines spend up to US$20 million per year on content.

Inflight internet is becoming more ubiquitous on aircraft now. This improves the inflight experience even more but comes at a big cost to the airline. Inevitably this cost will be passed on to the customer, like everything else.

If your IFE dies the next time you are flying, you might discover things that you have never seen before while you are twiddling your thumbs. If you are right next to the window, there is quite a lot to see. Many airliners have flat metal plates attached to the sides of the engines, known as nacelle chines, or strakes. Because engines are so large these days, at certain angles of attack, they block out some of the airflow to the wings. Strakes redirect the airflow by creating longitudinal vortices, which redirect the airflow onto the wings where it is required for lift.

If you look out at the tips of the wings on more modern aircraft, you will often see winglets sticking up. Given the number of different designs of these, you could be forgiven for thinking that they are there to make the aircraft look cooler, like mags on a car. They do in fact do this, but they have another purpose too. Winglets reduce wingtip vortices. The relatively higher pressure on the lower surface of the wing creates an airflow towards the wingtip, which then curls upward around it. When this airflow funnels out behind the airplane, a vortex is formed, which represents an energy loss. Winglets produce a good performance boost by reducing drag, which reduces fuel usage. The reduction in pressure and temperature inside each vortex causes water to condense and make the cores of the vortices visible. This effect is more common in more humid conditions.

If you are flying on a Boeing 787, you will see they don't have winglets, and given that these aircraft are the latest technology from Boeing, you might wonder what is going on. If you have a closer look, you will see that the ends of the wings curve rearwards. These are known as raked wingtips. These tips have a greater sweep angle than the rest of the wing, which have a similar effect on wingtip vortices as conventional winglets. They also offer weight savings over winglets, as they don't require an extra component to be added. Research by Boeing and NASA found that this design offers greater drag reduction (5.5 per cent) than traditional winglets (3.5–4.5 per cent).

One of the things that makes aircraft easy to spot are the 'contrails', or condensation (vapour) trails, behind each engine. But if you are looking at the winglets because the IFE is broken, you won't see your own contrails. The reason you won't see contrails from your aircraft is that it takes some time and distance for contrails to form after the vapour leaves the aircraft. Contrails usually form at higher altitudes, where the water vapour in aircraft engine exhausts combines with the low outside ambient temperatures to produce trails of ice crystals. They can be visible for a short time or may persist for hours and can spread to be several miles wide.

Conspiracy theorists reading this will no doubt be jumping up and down and tearing their hair out because they believe that contrails are actually 'chemtrails'. Those who subscribe to this theory speculate that the purpose of the chemical release may be, among other things, solar radiation management, weather modification, psychological manipulation and human population control. I'm no expert on these things, but what I can tell you is that there are no secret tanks on aircraft for holding these chemicals.

And contrails still happen in the middle of the Pacific Ocean, which would be a complete waste, don't you think? It all sounds a bit silly.

Well, that was a good little distraction from not being able to watch a movie, wasn't it? The good news is that the IFE just needed a reboot, and you can now resume watching your movie. We apologise if your choice of movie is not available on this flight. That will be because the good ones cost more, so we only supply B-grade movies. Dinner will be served shortly as well and, believe me, this will make your movie seem A-grade in comparison.

AIRLINE FOOD

By now you are probably hungry, and it won't be long before the cabin crew team are bringing you some sort of offering. Airlines have traditionally been gold medallists at screwing up airline food. In many cases, celebrity chefs are to blame for what you are about to receive, and my personal opinion is that celebrity chefs, and particularly those aligned with airlines, should be taken out the back and shot without a trial. If you are a celebrity chef and are reading this, you are no doubt currently a bit miffed. But bear with me while I explain, and if at the end, you still don't agree, then we can discuss your individual case out the back.

Some history is called for here to see how we got to where we are now. Airline food was all unheated meals in the early days. Then things started to improve during World War II when William Maxon invented the convection oven. As a result, airline meals could be precooked and heated quickly on board. Once onboard, the chilled food needs to be heated using the aircraft ovens. Each type of dish will have instructions for its reheating and preparation. For most economy meals, reheating simply takes place in the provided tray. More luxurious premium cabin meals will often be reheated in a provided tray and then transferred to other dishes for serving. These ovens used on an aircraft are specialised convection ovens with meals loaded on trays into the oven and heated using hot air, which usually takes around 20 minutes.

In the early years, airlines were prohibited from competing on price, so they competed on food. As unbelievable as this may seem, it has been suggested that at one point, United Airlines had the best food in the sky. In the 1970s and 1980s, airlines were deregulated in the USA so food became secondary to lower airfares. Then came loyalty programs, which encouraged people to stick with one airline no matter how horrid the food. As a result of these changes, we now have all the airlines in a race to the bottom.

Examples of this are many, but here are a couple. In the 1980s, American Airlines realised it could save $56,000 a year by removing a single olive from each passenger's salad. And recently a business-class passenger on a Japan Airlines flight to Japan from Jakarta who ordered a vegan breakfast was presented with a banana and a pair of chopsticks.

A known airline strategy is that the more horrible you make it to fly economy, the more businesspeople will demand to fly up front in business and first class. The airlines make huge money out of people flying up front, which is why they make economy as oppressive as possible.

All airline food comes from the same place, an industrial kitchen near the airport. The numbers are impressive. The Emirates flight catering centre in Dubai prepares up to 170,000 meals a day. Under 'internationally recognised standards' the food can be held in a chilled state for up to five days before loading. 'Internationally recognised standards' here means it's recognised internationally that this is totally unacceptable. Hot dishes are made in large industrial pans and decanted into plastic containers with foil lids before being blast-chilled to around 5°C in 90 minutes. They're then stacked in chilled metal boxes until taken onboard.

Airlines have been notoriously bad at food service for a long time. On some flights, the plastic-wrapped sandwiches and limp salads are pathetic offerings. The hot meals aren't much better. And the food that airlines serve tends to be high in ingredients that aren't good for you. This is because sugars and fat help preserve the taste and texture of the food.

As you go higher in an aircraft, the air pressure and humidity in the cabin reduce. This dry low-pressure air suppresses the sense of smell. Smell is vital in taste. In addition, background noise, such as the noise of the engines and air flowing past the aircraft, further affect how you taste. These noises can repress the ability to taste sweet and savoury foods by up to 30 per cent. So the next time you eat in the air, try putting in earplugs or listening to music to drown out the aircraft noises.

Apparently, the solution is umami, which is the 'fifth taste'. Humans taste only five things: salt, sweet, sour, bitter and umami. Umami is the savoury flavour associated with glutamate. In taste tests accompanied by loud noises, the perception of umami flavour was more intense than those conducted in silence, while loud noises weakened the awareness of sweet flavours.

Airlines that care have been working on using more umami in their meals, for example a British Airways shepherd's pie that has glutamate-rich seaweed in the crust. Lufthansa is now adding more high-umami ingredients, including tomato oils and tomato concentrate to its inflight offerings, and United Airlines uses ingredients such as tomatoes, glutamate-rich spinach and shell-fish to increase the umami taste.

Of course, there is more to a dining experience than just the taste of the food. If you pay for a first-class fare from New York to London, you could be up for $10,000 or more. Consider that you are still getting from A to B, just like an economy-class passenger, except that you will arrive a second before them. The grog is better quality in first class, but how much can you drink without losing the plot? And yes, you have got a bigger seat, but there's still no guarantee you won't be sitting next to a drug dealer in shorts and thongs. So that means you are paying the remaining $8,500 for your meal.

Admittedly the food will be of a higher standard, though it will still come to the aircraft in bulk dishes before being plated up for serving. But having decided that the meal is costing $8,500, you would also be right in expecting to eat what you wanted, when you wanted. Many airlines, however, will still only load four of the prawns, four of the steak, two of the fish, one vegan and one veggie meal for their 12 first-class customers. So, if you are sitting in the wrong place, you won't get your first choice. If you are last to be served, it will be vegan or veggie as no one really wants those meals.

This is unforgivable. I personally would not fly on an airline in business or first class that penny-pinched in that way. And it seems that passengers may have had enough of the airlines that do. In July 2023, a passenger on a United flight was so unruly over not getting his preferred meal choice that the plane had to be diverted. No, it wasn't to pick up another meal; it was to offload the enraged passenger. The peasants are revolting, and so they should.

Further down in economy, expectations are, of course, lower, so it is harder to disappoint people, but the airlines still manage to achieve it. They will offer two or three choices and then run out of the most popular one. So, when a flight attendant comes to you and says, 'Chicken or beef?' and you say, 'Beef', she will then say, 'Wrong. It's chicken.' And that's what you will be left with.

If you are flying economy on an airline where customers purchase a meal as an add-on before departure, then you can find yourself eating a meal in between two starving passengers with no meal, who are drooling and eyeing off your leftovers, even before you have finished eating. And the fact that you are using utensils made out of balsa wood or plastic to save weight is a hazard too, as these will break, and spray tomato paste onto your or your neighbour's white t-shirt. Plastic cutlery makes food taste worse too, according to scientists. This is because of a phenomenon known as 'sensation transference', which converts a negative visual sensation into an unpleasant flavour.

At best, economy food is edible. At worst, it's an overheated plastic tub of sludgy stew, overcooked rice or leathery meat and vegetables that have been boiled to death like your grandma used to do. There's only one way to ensure your food isn't dried out and disgusting. Choose the sauciest options. Stews and casseroles always taste better onboard than pasta, noodle and rice-based dishes, which are best avoided. They don't hold their texture when reheated and turn into a solid lump.

If you think this is bad, spare a thought for passengers on a recent British Airways flight from the Caribbean to London Heathrow. Because of a stuff-up with the original catering being left unrefrigerated for too long, it had to be thrown out. Local staff and some of the cabin crew in the Bahamas 'winged it' and did a KFC run. Passengers in all classes on the 12-hour flight were served a small portion of KFC to tide them over. No matter what you think of KFC, after that long not eating, it must have tasted okay.

If you are someone who likes to keep their hygiene standards up, bring disinfectant wipes on board. Fold-down trays are generally known to be the dirtiest surfaces on an aircraft, with more bacteria than a toilet seat. Sadly, these trays are cleaned far less than you would think. I might be stating the obvious here, but don't go running to the toilet to eat your meal off the seat. That's not fair on those who want to use the toilet as a toilet, so no, okay? Just no.

So, the inflight dining experience is not ideal, which brings me back to the celebrity chefs who latch onto airline teats, achieve nothing useful and inevitably add to the cost of everything. This is not a new phenomenon, but it surely all started because one celebrity chef had nude photos of an airline CEO, and once one airline was doing it, they all had to. There is some history of this.

Gordon Ramsay joined the International Culinary Panel for Singapore Airlines way back in 2006. Daniel Boulud has collaborated with Air France since 2016, creating menus for the business and first-class cabins. Neil Perry has worked with Qantas since 1997, designing the offerings for the first- and business-class cabins and lounges. Heston Blumenthal tackled British Airways' catering in a very experimental manner on *Heston's Mission Impossible* television series, unfortunately with limited success due to the unique challenges of the aircraft galley and cabin.

What can these people offer that a good cookbook and fresh ingredients can't provide? Are celebrity chefs a real solution to the airline catering predicament, or should the airlines be spending more money on their catering budget and their customer experience rather than buying into a celebrity chef's brand?

Gordon Ramsay has revealed he refuses to eat on planes and brings his own spread: 'I worked on airlines for ten years, so I know where this food has been and where it goes and how long it took before it got onboard.' He has a vested interest in knocking inflight catering now, because his restaurant, Plane Food, based in Heathrow's Terminal 5, sells takeaway food boxes designed to be eaten at 37,000 ft. For this hypocrisy, Gordon goes to the front of the queue out the back.

I like Nigella for obvious reasons, and Jamie Oliver is a man of the people and not at all arrogant. But celebrity chefs are all self-promoters of the highest order. And most don't even cook in their own kitchens. Imagine paying for a ten-course meal at one of Neil Perry's restaurants and walking out with a $400 per person hole in your credit card, while still being hungry. Meanwhile, the man himself is out self-promoting somewhere or sipping champagne on his yacht. If an airline captain left their name on the cockpit door, and put the minions in charge in order to stay home and self-promote, how would you feel about that?

As a side note, I've described the airline food space in less than 1,000 words, but you may be interested to know that a fellow called Richard Foss wrote a 248-page book on the subject in 2014 titled *Food in the Air and Space*. Who knows, it may be very good, but I'd be more likely to use it as a sleeping pill.

COFFEE

After dinner comes the coffee offering. I'm a coffee snob and I won't deny it. Liquid gold, a cup of joe, a brew, java, mud or whatever you like to call it, it's the source of life, at least for me. It's generally the first thing I do when I get up in the morning. I brew a cup and then sit and drink it, waiting for my heart to start. After a while, I can feel my body begin to function and gradually get up a head of steam. What would I do without coffee?

As someone who has travelled the world as a pilot, I've become something of an aficionado on the subject. In the air, at work, coffee is also really critical to your feeling of wellbeing and your level of alertness. Tragically, most airline coffee is rubbish. However, some of the better airlines use coffee pod machines for their first- and business-class passengers, which is passable.

For everyone else, the news is not so good. I've always suspected that coffee has a triple-layered life cycle. This can be the only explanation for the differing standards. In my theory, it is initially consumed in the classier coffee countries or regions such as Italy, Scandinavia, Australia, etc. The used grounds are then sent, exposed on the deck of a ship, to the USA. There, Starbucks and others reuse them as their primary offering, thus giving the stale insipid brew that Americans so love. These twice-used grounds are then sent to the airlines for a third life in the economy section, where they are mixed with sump oil to add some colour and flavour back in before serving.

It gets worse. Sometimes only decaf coffee is served, to make all the passengers more sleepy and less restless. This is a cabin crew initiative to ensure less work for themselves and can only be described as 'cruel but fair'. But wait, there's more. Water boils at lower temperatures as the cabin altitude climbs. At 2,000 m or 6,600 ft cabin altitude, the boiling temperature of water is 93.4°C, so if you are counting on the crew boiling the water to kill any nasties in it, think again. In addition, the water tanks are only cleaned out every quarter or so, and are filled in all sorts of different countries, so there is no guarantee that you won't get a bug from drinking airline tea and coffee.

When I was flying pre-Covid, coffee used to be brought to the flight deck in porcelain cups with a saucer. Shortly after I left, the change department decreed this a hazard and that coffee and other hot drinks for the pilots should henceforth be served in takeaway coffee cups with plastic tops on them. This was to avoid spilling the contents on the instrument panels and causing future delays.

But, as a result of Covid lay-offs, no one was ever given the job of ensuring consistent supply of these cups. Now it is becoming common for the pilots to be told that there are none available. This would mean no inflight coffee, which would be unaccept-able. So, when it occurs, pilots are delaying aircraft until cups are found. In some cases, staff are buying a stack of cups and tops off airport coffee shops just to keep the show on the road. The silliness continues and the romance has gone.

One of the great things about being international aircrew is that you get to sample the different types of coffee all over the world. In my early days in Qantas in about 1986, the crew had been out in San Francisco having quite a few beers and a meal somewhere, and on the way home, we stumbled (literally) onto an upmarket establishment that did fancy drinks. The captain led us in, and we all decided on an Irish coffee. After that, it seemed like a good idea to try a Mexican coffee. Then it seemed like a fantastic idea to go right through the coffee liqueur menu until we had tried them all. At the time, this was a barrel of laughs. Now, as I sit here writing at 3 am in 2023, I wish we hadn't done it as I haven't slept well since.

ALCOHOL AND DRUNKS

If coffee is not your thing and another drink is in order, there are some things you need to consider, because drinking at altitude is different than drinking on the ground.

The trouble is, our livers are magnificent multi-taskers, but they have their limits. If you push the boundaries by introducing more alcohol than your liver can handle, it's like inviting uninvited guests to a party in your bloodstream.

Picture this: as the alcohol ventures through your veins, it eventually arrives at its ultimate destination – the brain. Brace yourself, because once it reaches the brain, it transforms into a sneaky sedative. It slyly slows down transmissions and impulses between nerve cells, affecting your ability to think and move.

But here's where it gets intriguing. Alcohol is a master of duality. While it's technically a depressant, it has a hidden talent for removing inhibitions. It's like a roller-coaster ride for your emotions, unlocking a whirlwind of unexpected behaviour. Personally I can't see any downside here, but let's continue discussing it anyway.

The effects of this are worse on an aircraft. As previously mentioned, while soaring through the skies on your flight, the barometric pressure inside the aircraft's cabin differs from what we usually encounter on the ground. Think of it as being transported to a lofty mountain peak, where the air feels thinner at an altitude equivalent to 1,800 to 2,200 metres.

This lower pressure environment has an interesting effect on our bodies, hampering our ability to absorb oxygen, leading to a peculiar sensation known as hypoxia. Now, don't fret, as this is usually not a cause for concern. However, it can leave you feeling somewhat reminiscent of the aftermath of indulging in a few drinks, with a touch of light-headedness.

Therefore, if you drink alcoholic beverages during a flight, you may notice this light-headedness sooner, and so might the crew and other passengers if you drink too much. To be safe, you really shouldn't consume as much alcohol in flight as you would on the ground.

A complicating factor is that the air in an aircraft is very dry. So coupled with the diuretic effect of drinking alcohol, you may also become dehydrated much faster than you would on the ground.

So, the experts recommend to combat dehydration that you make sure you drink water with every alcoholic beverage. And minimise your intake of salty food, as this may have an adverse effect by making you thirstier and encouraging you to drink at a faster rate. Of course, we have just finished discussing how airline food is full of salt and umami, so as you can see, this is a vicious circle of trouble. Alcohol mixed with certain prescription drugs such as sleeping pills is on another level yet again, and should be avoided.

Cabin crew are quite within their rights to refuse further alcohol service to people who have had enough and are misbehaving, and they often will, so if you want another drink, don't be a dick. But sometimes cabin crew make the mistake of not cutting off alcohol in time to stop proceedings getting out of hand. On a British Airways flight heading to Saint Lucia in 2023, passengers claimed the cabin crew allowed a group of men to party and drink alcohol in excess. According to reports, a group of men were 'loudly partying' at the back of the plane for hours. The cabin crew allegedly allowed the men to have an 'endless supply of alcohol', the access to which was not curtailed until after the partying reached legendary status. This situation onboard the flight BA2159 from Gatwick escalated when a man was stabbed with a broken wine bottle. Witnesses also said the group of men were allowed to harass female passengers and block access to the bathrooms.

But bad behaviour isn't tolerated under normal circumstances on aircraft, and it's not okay to drink from your own duty-free alcohol on board. If you do this, crew will get very grumpy and possibly confiscate the grog until after landing.

There are other options for drinking on an aircraft apart from alcohol. You could always try water. Here's a tip: if you ever want to annoy your cabin crew or hold them up, ask for a plastic cup of Diet Coke. Because it fizzes up like Mt Vesuvius, these take longer to pour than any other drink. The cabin crew hate them and quite often tell passengers they have run out of it, rather than pour it.

INAPPROPRIATE BEHAVIOUR

What is deemed inappropriate these days is different to that which was deemed inappropriate 30 years ago. Inappropriate behaviour is taken much more seriously now and in many countries the incidences of it have reduced, mainly due to fear of the repercussions. The problem still exists in the international environment as some cultures still have a marked lack of respect for women.

Unfortunately, female cabin crew still have to bear the unwanted gestures, words and touching-up by male passengers from time to time. Alcohol does not help with this. During the bar service on one of our flights in the early 1990s, an incident occurred that was common at the time. Flight attendants back then each had their own bar cart to do the drinks service in economy class. As they finished up the bar service and were on their way back to the rear galley, a male passenger tried to touch up the leading flight attendant, but he missed. The leading flight attendant stopped a little further up to serve another drink, and the flight attendant behind had to stop right next to that same passenger.

This time he didn't miss, and he grabbed her between the legs. Her immediate thought was to punch him, but she resisted and verbally told him that this was not allowed. She said she would be reporting it. The passenger was initially spoken to by the customer service manager and then it was taken further to the captain.

The captain asked if the flight attendant wanted to press charges for sexual abuse. She said no on the proviso he would be dealt with on arrival into Sydney. The crew radioed ahead, and he spent most of his day being delayed by Customs, Immigration and Quarantine!

What's changed over the years is not so much the behaviour but more the tolerance towards it. These days, it's almost a given that the police will get involved.

Of course, it's not just passengers who are guilty of inappropriate behaviour. Cabin crew are exposed to it from within their own ranks as well. There is an infamous story of an incident years ago where three female flight attendants were sitting in the crew rest eating their meals, and the 'creepy' old senior flight attendant brought them a fresh fruit platter on a silver tray. The girls immediately noticed that he was holding the tray close to his midsection and his penis was in among the fruit. One of the girls quipped that the banana looked more useful, and they all told him to get lost.

It's no surprise that pilots are no different either. In June 2023, Ryanair's chief pilot was fired for 'a pattern of repeated inappropriate and unacceptable behaviour towards a number of female junior pilots'.

There have always been passengers, cabin crew and cockpit crew that have crossed the line and needed to be dealt with. The line between humour and harassment has moved drastically since I started at Qantas, and these days, anything vaguely inappropriate normally doesn't end well for the perpetrator.

ONBOARD SECURITY

On 12 September 2001, I woke up early in my hotel room in Cairns. I turned on CNN as I always did, and there was one of the Twin Towers collapsing in New York. At first, I thought it was a disaster movie, but as I watched, I realised what was happening. My initial thought was, 'This is interesting.' Watching throughout that day and the following days, I was filled with horror at the carnage and waste of lives. But even so, right then, I had no idea how that day would profoundly affect aviation going forward.

It was a new concept to use a commercial aircraft loaded with fuel as a weapon. What made it more dangerous was the willingness of the terrorists to die and become martyrs as well. This would change the whole concept of aviation security.

The immediate response of governments and security 'experts' around the world was to overreact. Governments need to be seen to be doing something, and security people need to start building bigger empires. Given the unprecedented events, this was completely understandable.

It did, however, take all the fun out of aviation. Cockpit visits were banned, airport security became a pain in the arse, and all of the romance of aviation died. My tolerance for troublemakers on my aircraft reduced to zero as well. Anyone who turned up drunk would be booted off, and anyone who threatened the crew or passengers would be handcuffed and handed over to authorities at the next destination.

Within a short time, we had super strong, bulletproof cockpit doors installed. Those cockpit doors became not only barriers to baddies trying to enter, but also barriers to interaction between the passengers and us. The tendency was to stay behind the locked door. This was partly because the old hijack rules of letting hijackers control proceedings were replaced with practices that emphasised protecting the cockpit at all costs. Terrorists could be executing people outside the door, but you would still not let them in. And if you were in the cabin when this happened, the other pilots wouldn't let you back in either. It sounds cold, but considering the destructive power of a martyr-flown aircraft, where the passengers would all die anyway, it was the only practical policy response.

Since 9/11, on each flight, the pilots decide on a password for the day, and they tell the cabin crew. Then when the cabin crew want to come to the flight deck, they need to give the password. This allows for some creativity and fun when coming up with the password. One of my favourites was 'Youdaman'. This meant the cabin crew would call up and say, 'We're coming up with your coffee and Youdaman.' To which I'd reply, 'Yes I am, and come on up.' Then they'd come and stand in front of the door. Once identified on our camera, we'd let them in. Of course, a password like that only worked if the cabin crew were fun as well.

So, security became the new deal and smart operators cashed in. The others who benefited were the previously unemployable, who got jobs at security checkpoints. These airport security officers are known for their severe attitude and lack of a sense of humour. On the face of it, they seem to lack any common-sense. 'Ah,' I hear you saying. 'You are being unfair here, Burfoot, because they are probably following rules and have no ability or permission to be flexible.' And I say, fair point, and thank you for bringing that to my attention, but no. The reason they don't have permission to be flexible is that their IQs are too low.

I've seen security people confiscate plastic water pistols and 10-mm spanners. What could you do with a 10-mm spanner? Remove the pilot's nuts? Unbolt the wings? Yet people are still allowed to carry duty-free grog onboard. Have you seen what a smashed glass bottle looks like when used as a weapon? An attack with a broken wine bottle was mentioned a couple of chapters back. Did you know that rubbing a plastic credit card on a hard surface will make the edge sharp enough to cut someone? Are you aware that fire extinguishers, freely available in the cabin, can be used as weapons? Do you get my point?

After 9/11, we started getting air marshals on board for certain 'random' flights. There would generally be two of them. I would get a call from security telling me where they would be seated. No one else on the crew would know. There was always one on the upper deck watching the cockpit door. The cabin crew knew we carried them randomly and got good at picking who they were. They'd come up and say, 'There's a "built" mid-thirties dude in smart formal gear who's not drinking in seat 18B. Is he a marshal?' I'd just smile. In my last few years flying, the air marshals seemed to have gone the way of the dinosaurs, though.

Disruptive passengers were still common post-9/11, even though it was well known that airlines and authorities were becoming way less tolerant of them. On one trip from London to Bangkok, the cabin crew advised that they had a guy downstairs who was being a dick. He was saying he was London Mafia, and he knew where we stayed in Bangkok. I went down to speak with him. He was short and fat and full of bluster, much like Boss Hogg from *The Dukes of Hazzard*. Bear in mind I'm over 6ft and very fit. I advised him that he had choices: leave the crew alone or be cuffed and met in Bangkok by the local police. He said to me, 'I know where you guys stay in Bangkok. I'll come and get you!' I pulled out a business card and said, 'Come to my room first. I'll be waiting for you.' Then I leaned into his ear and said, 'Now shut the fuck up.' And he did. These days, the captain wouldn't be able to go down. He'd have to send a more junior pilot. As I've said before, all the fun has gone out of aviation.

Of course, it's easy enough to take care of one nuisance, unless they are on drugs, but when a whole sports team starts playing up on their end-of-year party, you have a real issue on your hands. Not only do they encourage and embolden each other, but they are fit and strong as well. In the past this was not uncommon, and the only real option was to divert to the nearest airport and call the local constabulary.

Some of you will be disappointed to know that handcuffs were removed from aircraft many years ago as a result of them continuously going 'missing'. Nowadays if you misbehave you can expect to be cuffed with special double-sided cable ties. I don't recommend it. They look seriously uncomfortable.

A discussion about onboard security would not be complete without at least mentioning MH370. Everyone wants a pilot's opinion on what happened to MH370. I have no idea. If I did, I'd be writing about that instead. But it's worth watching the 2023 Netflix series, MH370: *The Plane That Disappeared*. This series gives a reasonable overview of all the different theories and ends up leaving you informed but even more confused than before. The series covers all the common theories and adds in some conspiracy theories. The problem is that none of them are compelling, because they either lack motive or commonsense says it would be impossible to keep them secret.

The most popular theory is a pilot suicide or a hijack that went wrong, with the end result that the aircraft ends up in the Indian Ocean west of Perth. The next theory is of a clever hijack that takes the aircraft northwest to land in one of the Stanley countries, Uzbekistan, Kazakhstan or Kyrgyzstan. According to the theory, this was carried out by the Russians along with the MH17 shootdown to punish Malaysia for something. There is no mention of what happens to the aircraft or the passengers under this scenario. Perhaps it was turned into a restaurant and the passengers are now working in a uranium mine.

Another theory is that the US air force shot down MH370 because it was carrying things to China that were not allowed to go there, such as sensitive technology. Supposedly under this scenario, the aircraft was shot down but remained completely intact and was spotted by a satellite on the ocean floor, impersonating a submarine, off the coast of southern Vietnam.

The theory that makes the most sense is: a pilot locks the other pilot out of the flight deck when he goes to use the bathroom. The remaining pilot then depressurises the aircraft, causing everyone to pass out from lack of oxygen. He has plans to do something bad with the aircraft to make a political point. But, somehow, his personal oxygen fails or runs out, and he passes out as well. The aircraft continues on course for seven hours, controlled only by the autopilot. Hypoxia is insidious. You don't notice it coming. I've been in an altitude chamber in the air force and experienced this. One minute you are fine, then the next thing you know, someone is putting a mask on you and asking if you are okay.

There is another slightly far-fetched possibility, which removes all evil intent from the event. It's a bit of a stretch but here goes: there is a fire in the vicinity of the flight deck and smoke is filling the cockpit. The pilots turn the aircraft around, using the auto-pilot heading selector to fly the heading they want while they try to deal with the problem. Fire starts to invade the flight deck, so the pilots leave and attempt to fight the fire through the cockpit door using portable fire extinguishers rushed from all over the aircraft. Eventually the fire is extinguished but the controls and switches that are used to direct the aircraft are melted. The aircraft continues for seven hours with the autopilot engaged until it runs out of fuel.

No one will really know what happened unless the aircraft is found, but suffice to say, inflight security is just as important as that which occurs before you get on the aircraft. Crew must remain vigilant at all times.

CELEBRITIES

One of the things that used to add a bit of interest to the flying game was having celebrities onboard. Pre-9/11, we'd offer them a visit to the flight deck. I met a few over the years including many Olympic athletes, international sports people and the like. I had the New Zealand rugby player Jonah Lomu come up to the flight deck once. When he shook my hand, his hand was nearly twice the size of mine. He was a gentle giant.

I also carried Elle Macpherson and Kylie Minogue, but neither was interested in being leered over by pilots, so they didn't visit. The King and Queen of Sweden seemed very nice. He came up. She didn't. My mother was on a few flights, and she came up quite often, mainly to make sure I was eating my greens and washing my hands properly.

Alan Jones, well-known Australian radio personality and one-time Wallabies rugby coach, came up once on a flight to Singapore. He had asked if he could come and listen to the State of Origin rugby league match live through our radios on the flight deck. We sat him in a seat at the back, shoved a headset on his head to shut him up, and left him to listen. At half-time in the game, he fell asleep. We continued to listen until the end. Just after the game finished, he was woken by a flight attendant who had come up to ask the score. The flight attendant said, 'Some passengers want to know the State of Origin score, skipper.' Before I could answer, Alan Jones piped up and said, '12:9', which was the score when he had fallen asleep. We all got a good laugh from that.

Post-9/11, everything changed. For a start, many celebrities decided it was safer to charter a jet. And of course, they couldn't come to the flight deck, so if you wanted to meet them you had to go down and bother them in the middle of their movie.

One time, on a flight from Los Angeles to Melbourne on the A380, I had the rapper 50 Cent and his entourage in first class. At the end of the flight, I left the cockpit to stand near the exit and say goodbye to the passengers. When 'Fiddy' came past, he said, 'Thanks, Captain,' and high-fived me. My hand is for sale. My contact details are at the end of the book.

THE MILE HIGH CLUB

Much has been said and written over the years about the 'Mile High Club'. Three million people in Denver, Colorado, and 22 million people in Mexico City think that membership is over-rated, because all of them live over a mile high. But technically, to join the club you have to have sex a mile or more above the ground, so it's back to the drawing board for those 25 million. And will any kind of sex cut it? It depends on who you ask.

The Mile High Club does exist insofar as if you satisfy the criteria, you are automatically a member, but that's about it. There's a mob on the Internet who call themselves the Mile High Club, but all they do is sell merchandise. So, if you want a t-shirt, certificate or branded luggage tag, you can go there and buy these. I suspect that flaunting your membership with these products might tell people more about your personality than your actual fame, so have a think first, please.

You can't actually apply for a membership to the Mile High Club. There is no form for it, and I suspect that the evidence required might raise eyebrows. But where there's a vacuum and a need, there's always someone ready to step up and fill it. And who better than me? I don't know about you, but I can feel a Nobel Peace Prize coming, or at least a knighthood for services to stupidity.

There does seem to be an intense and almost unhealthy fascination at the thought of becoming a member of the Mile High Club. For instance, when I am even older than I am now, and I am sitting on my porch in my rocking chair entertaining my grandchildren, I expect they will ask about it. 'Grandpa. Tell us about the time the couple in first class were sprung having sex in the toilet.'

'Awww, kids. Don't you want to hear about the time my 747 was upside-down, on fire and heading towards a mountain, and I saved the day?'

'Naah, Grandpa. Mile High. We love that story.'

'Okay. Wipe old Grandpa's chin, empty my bag, then bring me my pipe and I'll tell ya.'

In the good old days, there was plenty of hanky-panky taking place on aircraft. Cabin crew often congratulated a surprised couple after their raunch in a first-class toilet. While the quickie was happening, on-duty crew would receive phone calls in the galleys telling them to get up to the relevant toilet post-haste. On arrival they would wait for the door to open, hand the embarrassed couple a glass each of champagne and cheer!

There have also been many tales about first-class passengers having a go in the cabin and the crew having to throw a blanket over them. This still happens these days.

A cabin crew friend told me that during the breakfast service into a port in Japan on a 767, she was working in economy class on a cart from the rear galley moving forward on the right-hand side.

She was about to ask the Japanese passenger what he would like for breakfast when she noticed his new bride next to him was missing. Then she noticed the blanket over his legs appeared to be rhythmically moving up and down! She still proceeded without a fuss to offer him breakfast. He just muttered back, so she assumed he was not only busy but hungry at the same time. She left a breakfast tray for his bride. Her colleague on the other end of the cart showed her disgust once it was pointed out what was going on.

Since 9/11, we pilots haven't had much involvement in the club due to passengers being banned from the flight deck, but before that there was action aplenty. On my first trip in the company as a brand-new single-stripe second officer, I had just done a crew swap on the flight deck and was standing at the back. The cabin crew had just brought up a young mid-20s Australian nurse to see the flight deck. She wasn't hard on the eye, so I stayed and chatted with her for a bit. It was obvious she had consumed a few beverages and within a couple of minutes she said, 'I'm going to be chaperoned in Saudi Arabia for the next six months and I won't be getting any sex, so I want you, now, in there.' And she pointed to the crew rest.

I was flattered, of course, but being new, I had no idea of the protocol with these things. I was sure it would be a firing offence, so I declined. She was mightily pissed off. When I came back from my time off, the other pilots informed me that I was a loser and had blown the chance to become a legend on my first trip. Over the years prior to 9/11, I discovered that there were quite a few 'legends' among the pilots.

By far the most entertaining visitor I ever had on the flight deck was a young lady who was brought up during the daytime, shortly after I became a captain. She was drinking a glass of champagne and was quite drunk. She explained how she was on her honeymoon with her husband. They had been fighting, so she had come up to the flight deck to make him jealous. She then asked if anyone was interested in servicing her. Pilots are inherently good risk managers, so none of us were interested. Let me rephrase that. We were all interested, but all too smart to take that risk.

But one of the cabin crew guys later took her into the lower lobe galley beneath the main cabin floor and did the job there. While this was happening, the husband started to search the aircraft, including coming up and checking out the flight deck and crew rest. She was apparently nowhere on the aircraft, and this raised a bit of a concern for him. He wanted us to report her missing and turn back to look for her, much like a boat would if someone fell overboard. Thankfully after a little while, his wife surfaced, looking a little sheepish and smelling of sex. Another fight developed, and she admitted where she had been.

A few days later, I got a call from the manager of the 747. The incumbent at that time was a bit of a lad himself, maybe even a legend. He sent me an invitation to attend for tea and biscuits to explain. This was because the husband was threatening to sue the company and claiming his honeymoon had been ruined. So, in I went, and I told the boss the whole story. He chuckled and asked if she was hot. I said she was. Then he thanked me and sent me on my way. When confronted with the truth, the husband dropped the case. I hope they lived happily ever after.

Incidentally, the lower lobe galley on the old classic 747-200s was a known den of iniquity. It was on the same level as the cargo compartment, situated near the middle of the aircraft. Much of the food preparation happened down there and then food was sent up in carts via elevators. The galley was used by flight attendants from time to time to 'entertain' guests. If this was happening, there would be a do-not-disturb sign hung on the cart elevator door.

I could go on and on and on with stories about the Mile High Club, but that's not the purpose of this book. These days no one except you and me have a sense of humour anymore, and everyone gets offended too easily, so joining the Mile High Club onboard is no longer approved by management. This rule just serves as an encouragement, of course, and as long as there is plenty of alcohol around, there will be people attempting to join the club.

The toilet will always be the logical place for these festivities, particularly as there is a mirror in there to spice things up, but as you will see in the next section, it's a germ haven, so you will need to wear a full body condom. But at least it's not as bad for germs as the dining tray tables. One final warning. The cabin crew and I know how to open the toilet doors from the outside. Sure, it's an invasion of your privacy, but they'll still do it. I won't, though. You go hard.

AIRCRAFT TOILETS

There are those who like to spend a lot of time on a toilet contemplating or reading, and that's up to them. Personally, I like to get in and out in a hurry, and for that, I use oats. But aircraft toilets are a whole different story. I don't know anyone who thinks it's a good idea to go into one and read.

My doctor never went into one at all. Even on a 12-hour flight, he would not drink before or during the flight so that he could avoid aircraft toilets. He was a good doctor, so I took notice of what he said. But my doctor's advice notwithstanding, according to one source, passengers visit the loo on long-haul flights 2.4 times on average. This number makes sense. The first two visits are out of need, and the 0.4 is towards the end of the flight where they stick their head in, see the mess and change their minds.

Aircraft toilets have a lengthy history, and hopefully there are still improvements to come. Let's go back a bit and see where it all started. Aircraft as early as the 1919 Handley Page Type W and the 1921 DH.29 Doncaster installed basic toilets. The 1934 British Supermarine Stranraer flying boat had a *very* long drop toilet, which was open to the air below. Opening the toilet lid during the flight produced a whistling noise, which led to the aircraft's nickname, the 'Whistling Shit-house'. The Short Sunderland flying boat, which flew from 1938, was palatial in comparison, boasting a porcelain flush toilet.

Large bomber aircraft that flew in World War II, such as the British Lancaster and the American B-17 Flying Fortress, used chemical toilets (or buckets with seats). To use the toilets, aircraft crews would have to remove the multiple layers of warm clothing they wore to cope with freezing temperatures at high altitude. And if the pilot had to swerve suddenly without warning, which happened often, the crew might end up wearing the bucket's contents. For all of these reasons, bomber crews avoided using the toilets whenever possible, instead peeing into bottles or crapping into cardboard boxes that were ejected from the aircraft over enemy territory.

World War II fighter aircraft often used 'relief tubes', a funnel attached to a hose that poked out of the plane. Logic says that, given the cold temperatures, these devices would be useless unless the aircraft was being flown by a horse, but such some modern military aircraft still use them today.

For a time, aircraft manufacturers tried Anotec, the 'blue liquid' found in portaloos, to fight toilet germs. But if the tanks containing this liquid (and waste) leaked during the flight, as they sometimes did, large blocks of blue ice formed on the bottom of aircraft. These blocks would fall off and smash things on the ground during the descent, so Boeing and other manufacturers needed to find another option.

A clever dude called James Kemper patented the vacuum toilet in 1975. I hope he is very rich because he deserves it. These were first fitted to Boeing planes in 1982 and remain in use today. Kemper's new loo fixed the problem of spillage, to everyone's relief. The vaccum toilet uses a nonstick bowl that a blue substance called Skykem deodorises and disinfects. The toilet then uses vacuum suction to clear the contents. Adding to these benefits of vacuum toilets are that they are smaller, much lighter and use less water than previous designs.

The power of the vacuum system on these loos is legendary and the subject of much discussion at dinner parties. The vacuum is said to have a stronger suction than the 'Phoenix A' supermassive black hole. Airbus is reported to hold the speed record for sucking out waste, so their current toilets are known as the 'Formula 1's of aircraft toilets. By pushing the flush button on an Airbus, you are sending your poo on an exhilarating 200 kph journey through pipes to a holding tank in the back of the aircraft belly. There, it gets to party with everyone else's waste until sucked out by Winston the poo man at the end of the flight. Much as the pilots sometimes wish they could dump it on demand, this wouldn't be allowed and is not possible.

But back to the toilets themselves: there are rumours that if you flush while still sitting down, your innards will get sucked out and your head will cave in, but the chances are slim if you are slim. You would need to be big enough to form a complete seal around the whole bowl to achieve this, and there are only records of one unfortunate woman achieving it since the toilets were invented. She is still alive, but her head looks like a passionfruit.

At the end of a long-haul flight by an A380 or 747, there are, on average, 900 litres of waste for Winnie the poo man to remove. That's a lot of extra weight and is a significant cost for the airline.

Perhaps they need an incentive scheme for passengers to use the airport loos before their flight. You heard it here first.

If you are wondering what Winnie does with your poo once it is in his truck, he takes it to another area where waste can be dumped. At Auckland Airport this is called the 'Honeypot'. They must have processes for dealing with it after that, but the details would be beyond the scope of this book. You'll have to do your own research or contact Winnie direct.

Astute observers of aircraft toilets will notice that, even though smoking is banned on aircraft, toilets still feature ashtrays. The US Federal Aviation Administration actually still requires this, 'due to the fire risk caused by the possible disposal of illicitly consumed smoking materials in the toilet's waste-bin'.

There are always going to be smokers on aircraft, and some will think that they are cleverer than everybody else, so they will try to get away with a sly durry. The toilet is the most obvious place to try. But this reminds me of when I was in Venice once with a mate, late at night after a few drinks, and he announced that he was going to steal a gondola and take it for a ride. I watched from a nearby bridge as he futzed around in the dark for a while and finally, he yelled, 'They are all chained up.' Whereupon I said, 'Really? I guess you aren't the first drunken tourist to try that then.'

Fires in aircraft toilets are a big risk because it has more time to grow before someone notices it and sounds the alarm, and several crashes and other incidents have been linked to such fires. Because of this, toilet waste bins have bottle fire extinguishers attached to them, and the smoke detectors in the toilets will trigger alarms and flashing warning lights in the cabin and cockpit. It is actually quite rare to go more than one or two long-haul flights without someone setting off a toilet smoke detector.

So, having mentioned the limitations of the current aircraft toilets, it would be fair for you to ask, 'Well, what's the answer?' And what good would I be as a correspondent if I didn't have the answer? There is good news from the front on this. In 2016, Boeing released a new, larger, self-cleaning toilet design. It uses ultraviolet light to reportedly kill 99.99 per cent of germs (and remove odours) on the toilet, taps and counter tops. What's not to like about that? This fancy new crapper also has sensor taps, soap dispensers, trash lids and a hand dryer. Boeing is also working on a hands-free door latch and a vacuum-vent system, so you should forward any ideas for this to them – after you apply for a patent. This is all good news, and one hopes that they become main-stream yesterday.

The last word on toileting has to go to an explanation for paper cups in toilets. Some say these are for drinking tap water. That's far-fetched but possible I guess. Thinking people will tell you that they are for men with that early morning erection that just won't go way. Unable to point it downwards, they can hold the cup upside down above the toilet and pee up into it, thus avoiding making a mess.

TIME ZONE CHANGES

Time zones have managed to confuse many people over the years and still do. If you look at what's written on the Internet, there's a lot of gobbledygook and scientific speak. So as an extra service in this book, here is Captain Jeremy's simple version.

If you can accept that the sun rises in the east and sets in the west, you are off to a good start. That means that, due to the direction of the Earth's rotation, the sun effectively travels in a westerly direction around the world. As a result, the sun can't be up in all places at once, hence the need for lights.

Humans base their day, on average, around the sun being at its highest point around noon. But the sun moves westward at 15 degrees of longitude per hour. Therefore, each 15 degrees of longitude to the west requires its own time zone to allow the sun to be highest at around noon in that zone.

In establishing the chart of time zones, for one reason or another it was based around the town of Greenwich in the United Kingdom. This is known as the Greenwich Meridian and is 0 degrees longitude. The time there is known as Greenwich Mean Time or GMT.

If you look at a diagram of time zones on a map (below), The Greenwich Meridian should be in the middle. To the east (right), and approximately every 15 degrees of longitude, the time zones become GMT+1, then GMT+ 2 all the way out to New Zealand near 180 degrees east and GMT+ 12.

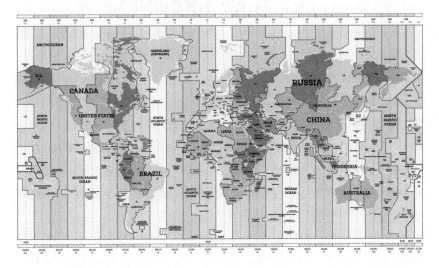

It's important to remember that GMT+12 is always ahead of all other times zones and all time zones are ahead of GMT-12.

Let's do an example to see how this can affect you. New York is five hours behind London, and you are flying New York to London. If the flight is 7 hours long and London is 5 hours ahead of New York, then when you get there, it will have effectively taken 12 hours (7 + 5 = 12). If you left at 8 am New York time, you would arrive at 8pm London time.

Going the other way, if you left London at 8am, you would arrive in New York at 10am, having effectively only taken 2 hours (7 − 5 = 2). When the Concorde was flying this route and took only 4 hours of flight time, they would actually get there at 7 am New York time, 1 hour before they left $(4 - 5 = -1)$.

The rule is that if a place is east of you (or to the right on the time zone map), they are ahead of you in time. If it is to the west of you (or left on the map), they are behind you in time.

The real problem with time zones arises when you start to work out what day it is in any place. The farthest east you can go is 180 degrees, so there, in the middle of the Pacific Ocean, they have plonked an International Dateline. If it's the 24th in New Zealand, it's the 23rd just to the east across the dateline.

So, let's do an example of flying across the dateline. You leave Sydney at noon flying to Los Angeles. Sydney is GMT+10 and Los Angeles is GMT-8. So at noon in Sydney on the 24th, it's 6 pm on the 23rd in Los Angeles. But the flight is 12 hours long. The time zones are −18 apart. The maths is $(-18 + 12 = -6)$. So, your flight would arrive in Los Angeles at 6 am on the 24th. This is effectively arriving 6 hours before you left.

Going the other way, if you left Los Angeles at 11 pm on the 24th, the maths would be $(+18 + 12 = +30)$ so you would arrive in Sydney at 5 am on the 26th and you would have effectively lost a whole day.

Crossing the international dateline raises some interesting possibilities. If you leave Sydney on the 24th and it's your birthday, you can have another birthday when you arrive in Los Angeles as it will be the 24th again. But if you leave Los Angeles on the 23rd, you will arrive in Sydney having missed your birthday. I've missed a few Christmases and birthdays this way, which explains why I look so young.

The best use of the dateline I've heard of is the cruises that leave Auckland, New Zealand, on 30 December. These cruises go out and sit 100 m west of the dateline and are the first to see in the new year in the world. The next day, hangover cures are all the rage in the morning and then the ship cruises 200 m to the east and has another new year's party on the other side of the dateline. What's not to like about this?

STEP CLIMBS

During a standard long-haul flight, an aircraft should climb every now and then to a higher level as its weight reduces with fuel burn-off. If you are smart, you won't notice because you'll be wearing earplugs to make the food taste better. Otherwise, you'll hear the engine noise increase for a few minutes and feel the nose rise a bit. Then the nose will lower again, and the engine noise will reduce once more. This is a step climb.

The reason for this step climb is that as the aircraft moves to greater heights, the air becomes less dense, which also lowers the drag, while the air pressure and temperature both decrease. This reduces the maximum power or thrust available to an aircraft's engines. An aircraft aims to cruise at the maximum altitude it can generate sufficient lift to maintain, which minimises its fuel consumption. But as the aircraft's weight decreases due to fuel burn, it must adjust its position to remain at the best cruising altitude.

Fuel burn can be significant on large aircraft. On an A380 it averages out at around 12 t per hour, while a 747 is around 10 t per hour. A 787, which carries way more than half a 747 load, is considerably lower at around 5 t per hour. This explains why airlines are opting for 787s and A350s over the fuel-guzzling four-engine aircraft.

In the early days, most pilots used the more efficient simple cruise climb, a slow but non-stop journey to the final cruising altitude. But an increase in air traffic has mostly made it too dangerous to use the simple cruise climb, which is where the less efficient step climb, usually around 2,000 ft, comes in.

In areas of very high air traffic operations, even step climbs may not be available when pilots want them. You have to plan in advance for this by carrying more fuel. Early in 2000, I found myself on a trip to Frankfurt with a permanent first officer who I'll call Dufus. Permanent first officers have usually failed command training twice or have been told by the company not to bother, because their overall standard is too low to become a captain. Dufus was pleasant enough but just a bit slow.

We took off from Singapore, heading to Frankfurt in the middle of the night. Dufus took the first two hours off, then came back on duty as we were approaching Calcutta (now Kolkata). As I was about to head off, I briefed Dufus, saying, 'Every level is full of aircraft. There is no way you'll get a climb until aircraft start diverging after Afghanistan, so relax.' Then I went off for a few hours.

I was called back sooner than I thought and figured something was up, so when I got on the flight deck, as a joke, I said, 'What's up? Are you lost?' No one laughed. Dufus had a chart open, covering all the screens, and didn't look up. The second officer, sitting in my seat, looked at me and nodded his head. He got out and let me in the seat. Then he briefed me.

After I had told Dufus he wouldn't get a climb, he had asked for one anyway. The Indian controllers had offered him a diversion to the south to get a climb, which he had accepted, but what hadn't crossed his mind was that there was no space at the higher level for him to get back on track. So instead of heading up over Afghanistan and Russia, we were now over Iran.

I sent Dufus back to the bunk, and the second officer and I got stuck into sorting out the mess. In those days, no one was flying over Iran. The Iranians were very grumpy about us being there and wanted to know our planned track through Iran, and if we would like a Shahab-3 missile up our tail pipe. Every time we changed controllers, they asked again for all the information, as they didn't have our flight plan. Finally, I managed to make a plan using the chart, and we kept plugging in new waypoints into the navigation system. As we passed over Tehran, the second officer took a fuel reading and then contacted the company on the high-frequency radio to ask for a new flight plan from Tehran to Frankfurt, to see if we could make it.

Eventually, we got spat out the other end of Iran into Turkey (now Türkiye), where the controllers gave 'less of a shit', and they were more helpful. Then the flight plan turned up, and it was touch and go for Frankfurt. Sadly, it wasn't going to work on this occasion, so we landed in Vienna, Austria, for more fuel. Then we flew on to Frankfurt.

On the ground in Frankfurt, Dufus said to me, 'Are you going to tell the company?'

I said, 'No. I think you've learned enough and suffered enough already, but if they ask, I'll have to.' Surprisingly, even though we were over Iran without an explanation, and in spite of a diversion to Vienna and being three hours late into Frankfurt, no one ever asked. I hadn't seen Dufus for a while when I retired, so I assume he wandered off and forgot how to get to work.

FOUR

BEHIND THE CURTAIN

YOU'VE BEEN PEEKING AGAIN, HAVEN'T YOU?

WHAT PILOTS DO ON THE FLIGHT DECK

So, apart from getting lost, what exactly goes on behind that bulletproof cockpit door while you are eating umami and drinking too much? People are always asking me, 'What do you pilots actually do? The autopilot does it all for you, doesn't it?' I normally say, 'Yes. Thank God, someone has finally figured it out. The pressure of keeping it bottled up was getting to me. I drink coffee and eat food and get paid a shitload. Case closed.'

But you are smarter than that, and you know the company wouldn't pay me unless I was worth it. Economics 101 would eventually find me out of a job or working for minimum wage. So, what do we do then?

As a captain, you are the CEO of the whole operation. You have sole charge of the aircraft and its operation. While you are in the air, you don't have to follow the instructions of your company or air traffic control if you decide it would compromise the safety of the aircraft and its occupants. Because of this relative autonomy, you have to be able to defend your decisions in a court of law after the fact if something goes wrong. It's all on you, so it needs to be taken seriously.

The captain sets the tone for everything that happens onboard as well. A good captain encourages a relaxed professional environment with good communications. You lead by example and your authority, although written, is best earned from your leadership style. A reasonable indicator of how well you are doing this is whether the crew call you Captain or Skipper, where 'Skipper' means they respect you and like you. The captain's job is to get the best from their crew so that the overall execution of the flight is carried out to a high standard.

First officers are deputy captains. Their job is to back up the captain, make inputs in decision-making and assume command of the flight deck when the captain is in the crew rest or cabin. A good first officer will always be second-guessing the captain: if they do something unexpected, the first officer should question it. This 'two heads are better than one' concept is why there still needs to be two pilots on the flight deck for the foreseeable future.

First officers also need to be diplomats. Even if they know more than the captain, fly better than the captain and get more attention from the flight attendants than the captain does, they need to somehow not make it look that way. This takes skill. A good captain will recognise it, admire it and chuckle quietly to themselves.

Second officers (S/Os), when carried on longer flights, are primarily there for inflight relief, which means to sit in a cockpit seat when the captain or first officer is resting. But there is more to their job than that. Right from the moment they turn up, they are used to enhance the operation.

They start by helping with the flight planning and fuel decisions, and filling out the required paperwork. When the crew gets to the aircraft, the S/O gets anything out that the crew might need for the pre-flight and departure. Then the S/O conducts a walk-around inspection of the outside of the aircraft, a proverbial (but not literal) kicking of tires, so to speak. Usually, this would already have been done by a ground engineer, but, especially overseas, it is common for S/Os to pick up things that have been missed. This could include the likes of tyre wear limits, brake wear limits, missing static discharge wicks, oil or hydraulic system leaks, etc.

When we had the 'Combi' 747s, with a freight area at the rear of the main deck, it was the S/O's job to check that as well. They were also the designated firefighter for that section, requiring extra fire training each year.

After returning to the cockpit after the outside inspection, the S/O calculates take-off performance to compare with the captain's and first officer's calculations. Sometime towards the end of the pre-flight, the refueller will turn up, and the S/O checks that the correct amount has been loaded and then signs for it. After refuelling, the S/O does a radio check on the high-frequency (HF) radio, and if they haven't already done so, they check the technical log for any maintenance items that require different pre-flight or inflight procedures. A good S/O never has to be asked for anything. They will be one step ahead of the captain and first officer the whole time.

After that, the S/O's job is to monitor the whole operation until they are released on break or asked to occupy one of the pilot's seats. When in the seat, the S/O usually takes over the radio work and log keeping. In the good old days, they'd get weather updates on the HF radio and load navigation waypoints, but that's all done electronically in modern times.

In the 1980s and 1990s, anyone trying to communicate on HF radio had a tough time. I remember it well and still occasionally wake up screaming, 'Bombay, Bombay' in the middle of the night. Going from Southeast Asia to Europe involved crossing the Indian subcontinent and the surrounding seas, and the only form of communication with air traffic control was by HF radio, which is noisy, unclear and just plain hard work.

The air traffic control units in this area were all linked and used the same HF frequencies. All aircraft in the area had to report at every waypoint that they crossed. So, imagine, there could be over 200 aircraft trying to get position reports in every couple of minutes. The first thing to establish was which HF frequency they were using of the five or six written on the charts.

The general rule was that if you were further away, then use a higher frequency and the lower the sun, the lower the frequency. So, when asked to call Karachi on a frequency, you'd dial up the frequency. If you heard nothing, then you had to hunt around the alternate frequencies. You'd know when you'd found the right one, because there would be ten people all talking over one another. To get a word in, you needed to be aggressive. Frustration was always evident. A common sort of exchange would be an Indian controller saying something like, 'Qantas one, Qantas one, Bombay, go ahead, go ahead . . . shut up Delhi. Qantas one, go ahead.' It was a shit fight and I hated it.

Second officers on most airlines are trained to land the aircraft in the simulator but are not allowed to do this in the actual aircraft. Their licence is for inflight relief only. It's nice to have a second officer on a flight, because if you don't have one, then all these tasks they do are shared between the captain and first officer. And when you are tired, it's nice to have a third set of eyes watching and giving reminders when needed.

Second officers have always jokingly been known as 'The captain's sexual advisor' as well, meaning *If the captain wants his fucking advice, he'll ask for it*. That's a humorous hangover from the old days when speaking up on the flight deck wasn't encouraged. Thanks to crew resource management (CRM) training, everyone is encouraged to speak up these days, and safety is enhanced as a result. Even so, a cockpit crew is not a committee, and the captain, who has all the responsibility, always has the final say.

Right about now, I'm guessing you are thinking my response to the question of what we do on the flight deck has been much like a politician who gets up to announce that he really doesn't have anything to say. Well, I've covered the pre-flight prep, and if you want more detail covering the rest of the flight, read on.

As mentioned earlier in the book, we manage energy, which, given the amounts involved, is a big deal on a large aircraft. It directly reflects on the airline's bottom line, management bonuses and, ultimately, on ticket prices. We also monitor risk and actively sniff around for it. As well as that, we liaise with the cabin crew as to the best times for cabin service given the known forecast weather and turbulence en route.

We report to and liaise with air traffic control. We continuously monitor the destination weather, and that of alternate airports along our flight path. We divert around areas of thunderstorms and turbulence so that your tea is not spilled any more than it needs to be. These diversions require permission from air traffic control.

We manage the crew roster. We keep the company appraised of progress. We enter any faults in the technical log and advise engineering at the destination. We monitor aircraft tracking accuracy. We plan en-route climbs and monitor fuel usage. We make PAs as required and also to reassure people that we are still there and haven't been left behind at the airport gate. We do full cockpit and systems scans periodically to spot any problems before they develop.

On international flights we study specific en-route and destination procedures in online manuals. There is an immense amount of local knowledge required for international pilots to do their job well because of all the differences in procedures between the various countries that you fly over. Thanks to ICAO (the International Civil Aviation Organization), many rules and procedures are standardised, but many countries still insist on their different procedures as well.

Pilots need to know all the rules, regulations and differences for all the countries they fly into. It would be impossible to remember them all, so most pilots have notebooks with reminders for each country. The concept of safety in aviation relies on the rules being followed. It's not a choice thing. Many of the rules and regulations came into being as a result of the learnings from times when things went horribly wrong. If you break the rules, you'd better be able to defend your actions in a court of law, as people's lives are at stake.

In some countries, if you crash or have an incident in an airliner, the authorities will immediately throw the pilots in the slammer, pending an investigation. In 2022, a Latam A320neo was hit by a fire truck on a runway in Lima, Peru. The accident killed two firefighters and injured a third but didn't result in any injuries on the aircraft. In the immediate aftermath, police hauled the two pilots off to jail, where they were held for almost two days.

In countries where a fair trial is unlikely due to corruption and cover-ups, pilots know that the first thing they need to do after ensuring everyone's safety is to get away and hide. Reasonable airport authorities who don't agree with the laws will count to 100 while you do this. It will be then up to the airline and their pilot's union to get them out of that country. It's real cloak-and-dagger stuff.

Using ICAO rules as the master rule book, civil aviation organisations in various countries create their own rules, which they push down to the airlines. The airlines then put out their own rule book for pilots to follow, which has to be at least equal to the government regulations, or more strict.

These company procedures are known as Standard Operating Procedures (SOPs). SOPs are a great concept for safety. When everyone is dancing to the same tune, everyone knows what to expect and safety is enhanced. If someone does something non-standard, it will instantly be recognised by the others and questioned. Airlines where pilots stick strictly to SOPs generally have a better safety record.

Of course, any of these procedures can be changed at any time, and they relentlessly are. A little-known fact is that airlines, and indeed most large corporates, have a change department. No one knows where it is located, but it surely exists. These departments are sometimes known as the Ministry for Evil Laughs. Their mission statement is 'The world needs more mwah ha haa', and their job is to keep people on their toes. Successful change departments have reverse-engineered Spencer Johnson's excellent book *Who Moved My Cheese* to come up with an algorithm for change that will be the most disruptive to the end user. The goal is to wait until people have just got used to the last change, then implement the next one. There is no requirement for the change to be an improvement, as long as you say it's world's best practice, whatever that is. In this way, the change department keeps everyone on their toes and has a lot of fun doing it.

English is the international language for aviation, but many countries don't speak it very well and all have accents (except us). In Mexico they speak English to English-speaking pilots and Spanish to everyone else. This makes everything useless to a point, because we listen to all air traffic control conversations to become aware of possible conflicts. When Spanish is being spoken, it just sounds like an endless conversation about ordering at an Italian restaurant.

Foreign accents get easier to understand with experience. Pilots new to the international flying game would listen to a controller and then frequently look at me and say, 'What did he say?' It would be easier for me to reply to the controller rather than explain it to the junior pilot. Having done that, I would turn to the junior pilot and say something like, 'He said that he met your mother once in a bar in Hong Kong, and he was wondering if you were his son.'

As well as accents, in many places, the correct wording for communications with air traffic control is different. In some countries, air traffic control expectations are different too. For example, normally an invitation to descend to a lower altitude is just that, and this can be done at the pilot's discretion, whereas, in New York, if they ask you to descend to a specific altitude, they want you there yesterday, and best you get on with it or they'll remove you from the queue and slot you back in later when it suits them. As well as this, every foreign airport has different surrounding terrain, different suitable alternate airports and different ground anti-icing procedures.

Incidentally, the New York controllers are the grumpiest I ever encountered anywhere in the world. I'm not sure why, but I bet it's a source of pride for them – as it should be, because they do it well. One particular controller got more than he bargained for when he got grumpy with me once. We ended up having a 'grump-off', which I believe ended in a draw when I suggested over the radio that being grumpy with Captain Grumps himself would achieve nothing, and that perhaps it was time for him to take a holiday. He was much friendlier after that.

It also pays for pilots to know what to expect in a descent and arrival clearance. For example, going into London, they will clear you to descend, which you would do at your normal point, if you didn't know better, but then after you start the descent, they will ask you to be at flight level 250 or below passing a certain point, which now means you are seriously high, so out comes the speed brake and down goes the nose.

Knowledge of en-route requirements in areas of high traffic concentrations are also important. Many flights leaving multiple departure points in Asia pass through Afghanistan's airspace, which has limited capacity to support air traffic. These flights must apply for an arrival time over the Kabul FIR (Flight Information Region) boundary. The Bay of Bengal Cooperative Air Traffic Flow Management System, or BOBCAT, then allocates entry slots into the Kabul FIR to each aircraft before it departs. The aircraft crew then have to plan to depart at a time that allows them to reach the edge of the Kabul FIR at the correct time. Are you still with me? I hope so because I purposely typed slowly to allow you more time to absorb the detail.

The more pressing point with these flights to Europe leaving in the evening is carrying the right amount of fuel. With bottlenecks like the Kabul FIR, it doesn't matter how efficient the countries are before you get there, you still have to be correctly spaced for when you get to the limiting area. So, heading over the Bay of Bengal and then over India, every spot is usually packed at all cruising levels (take note, Dufus). It's not uncommon to have to fly at a much lower altitude than you would like, which is very inefficient and costs in fuel burn and hence money.

On the other hand, carrying too much fuel just in case is also inefficient and costly, so it's a balancing act, and this is where experience pays big dividends. The trick is to get to your destination as efficiently as possible before all the fuel is replaced by air in the fuel tanks, and everything suddenly goes quiet.

In these areas of high air traffic, we always keep a good eye on the outside. You can generally see the aircraft above and below you, and you will also see those aircraft coming from the other direction at some point as well. Quite often you will have another aircraft offset slightly to one side, which passengers will be able to see.

One of the finest theories I have ever heard about nearby aircraft was just after 9/11, when I was a passenger between Sydney and Auckland, sitting in an aisle seat. Sitting in the window seat was a young lady, who pointed out to the young man occupying the middle seat that there was another aircraft below us. The young man decided to impress her by offering an explanation for this. He said, 'Since 9/11, aircraft fly in pairs so they can help each other in case of trouble.' I decided to be helpful by asking how they would help each other. He struggled with this for a bit, so I asked if perhaps they had 2,000-ft-long rope ladders for rescuing people from hijacked aircraft. He supposed this could possibly be the case. I then shut up and let him get on with it. He was on a mission, after all, and that's admirable.

Pilots also communicate with other aircraft about turbulence and estimated times at certain points and, sometimes, just for fun. On one such occasion we were heading southeast over the Bay of Bengal with our destination as Singapore. Ahead of us by about 10 km, but 2,000 ft above, was a British Airways 747. I called them up and said, 'Hey Speedbird (the British Airways callsign). How about making a few turns from side to side and we'll get some photos and send them to you?' They thought this was a really good plan and started weaving from side to side. Meanwhile, we sped up and slipped underneath them and slightly ahead.

When it came time to descend into Singapore, we were in their way, so they had to be diverted off track to the right. We ended up landing five minutes before them. Later, in the bar frequented by British and Qantas crews, their captain came over and asked if he could get hold of the photos. We told him there weren't any. We had just been winding him up. To his credit, he thought it was a hell of a joke and bought us a beer. I've always liked the British.

We can't finish talking about what pilots do on the flight deck without talking about eating, and specifically crew meals. Pilots generally do not eat passenger meals. They may get offered leftover first or business class food if the flight attendants like them, but there are always separate crew meals loaded. These are much like business class meals but delivered in the same configuration as an economy tray meal.

Because it's not ideal if all the pilots go down with food poisoning together, the captain and first officer must eat different meals.

There are also commonsense rules on some of the types of food that are more likely to give upset stomachs, and eating these is discouraged but not banned.

In general, pilots are kept well topped up with many of the wrong types of calories such as chocolates and cookies. If you have a sweet tooth and poor self-control, this can be disastrous. That is why many pilots are oversized and the others spend a lot of time in the gym.

WHAT PILOTS TALK ABOUT ON THE FLIGHT DECK

Having discussed what pilots do on the flight deck, I'm going to come clean and say that even with all the things we do, longer flights give us a lot of spare time. Given that about two thirds of my flying hours were at night, strategies needed to be put in place to keep awake, and this is where conversation between the pilots comes to the fore. So, what do we talk about?

Before we move onto topics, one thing that pilots need to consider is that the boss is always listening. A cockpit voice recorder (CVR) is used to record the audio environment in the flight deck for the purpose of accident and incident investigation. The CVR records and stores audio signals from the microphones and earphones of the pilots' headsets, and from an area microphone installed in the cockpit. It records for about two hours then starts recording over the old recordings, so that the last two hours are always available.

There have been numerous examples of CVR recordings being used for purposes other than accident or serious incident investigation, including publication in the media and listening to flight crew conversations by airline staff for internal use. The CVRs have an erase button on the flight deck, which pilots will normally push at the end of the flight. Airlines have always tried to get rid of the button, and unions have always insisted on their right to privacy after a successful flight. Pilots are aware of the CVR, but it doesn't seem to limit our discussion topics.

So what do we talk about? First up for discussion is usually what you've been up to in your life, and if you have any plans for the days off in your next destination. This leads into relationships and often pilots find themselves discussing and offering advice on this.

One of the common themes of the younger guys I spoke with was that they were thinking of getting married but weren't sure. They wanted an opinion, so they'd ask me. The first thing I would say was, 'If you have to ask me then the answer is no. It's hard enough when you are sure. If you aren't sure, it's a nightmare.' They would contemplate my sage advice for a minute, and then I'd completely throw them by saying, 'But on second thoughts, I think you should do it.' They would then look at me and ask why. And I'd say, 'Why should you be the only one that's single and happy?'

The next theme for discussion is what is going on between the pilot's union and the company. There is generally a stoush in play or one to look forward to in the near future. Everyone seems to have an opinion on industrial matters and these opinions vary. Surprisingly, self-interest seems to drive the differences in opinions. I enjoyed pointing this out to people. One particular junior pilot was frothing at the mouth to me about senior pilots having too many benefits, and this pilot had a solution that would fix the problem. As we had a seniority system in place, I suggested we could bring in his plan from his seniority number and below. After a minute's thought, he decided that wouldn't work. I said, 'Why? Because you have nothing to gain?'

But despite differences in opinion between individual pilots, there was unity in the fight against the company. The general theme is that the company wants to reduce pay and conditions to increase profits and management bonuses. The airlines did not invent this theme, but they have perfected and enthusiastically pursued it.

At the start of a wage bargaining round, the pilot representatives report back to the membership that the company has offered a pineapple, along with instructions about where to insert it. This makes everyone dig in a bit more.

After a few months of silliness, the company then offers a watermelon, which is harder to insert but not as rough. This too is rejected, and meanwhile the company keeps all the pay-raise money interest free. Around this time the senior management will take a pay raise for themselves at least twice the size of what the pilots are after, while the company also has plans to spend $100 million defending against the pilot's bid for a $50 million total raise. All in all, there is more skulduggery and intrigue in the process than in an election of a new Pope.

It needs to be mentioned that senior airline management don't tend to like pilots because they know that pilots can see through their bullshit. This always causes a conundrum at the bargaining table for wage rounds. The management will put forward what they want to pay the pilots in rupees or Turkish lira, and how many conditions they want to scrub. The pilots will look at their offer and say, 'If we are silly enough to accept your offer, then we aren't smart enough to fly your planes.'

I was on good terms with one of the very senior managers, whose name you would recognise, and we would have friendly discussions about how things were going operationally. He was genuinely interested in my opinion, which I appreciated. On one occasion, bargaining was happening between the union and the company, and the company had threatened to bring in pilots from overseas if our pilots refused to work for a bowl of cornflakes and a 'Happy Meal toy'. I said to my friend, 'What's with that? There are no pilots available overseas and it would blow the company's credibility out of the water.' He responded with, 'But you have to have a plan B.'

I said, 'But if the plan B isn't viable, then it's not a plan B.' This didn't seem to bother him.

Because nothing makes a bored pilot happier than having a good whinge, rostering and scheduling are often discussed on flights too. On this topic, self-interest is again alive and well. It's worth discussing crew scheduling here, as it's a topic of great importance in any airline.

Airlines generally always have spare crew available in case of sickness or disruption to a service. Occasionally they don't, and this is why airlines sometimes cancel flights due to crewing issues, or hold up their departure waiting for replacement crew. Given the size of some airlines, this can be a complicated under-taking requiring a lot of computing power and some outstanding operators.

The relationship between crew and scheduling can some-times be fractious, because crew don't always want to work, and schedulers have a job to do to keep the airline running. There are advantages to be had from knowing the agreed rules well. Sadly, this includes knowing when scheduling is stretching the truth about what you are required to do. That is generally the exception rather than the rule, but it happens to everyone at some point. It's not too much of an issue for young single folks, as going to work is fun, particularly if the destination is good.

Qantas has a seniority system, as do most airlines. Scheduling systems in airlines worldwide are very similar. In the case of Qantas, your working schedule is divided into 56-day rosters. Every fourth or fifth 56-day roster is a 'blank line' with no scheduled flying. For 56 days, you are available to the company, as needed, for flights or standby duties.

When you're on standby duties, you are paid to be ready to go at short notice if the company needs you. The company's perception of this is that you are sitting in the car with the engine running, garage door open, in full uniform (including hat) and ready to burn some serious rubber when the phone rings. The pilot's perspective is that you are playing golf in a resort in Cancun with your mobile phone on vibrate only. The truth is somewhere in between, and you need to be ready to fly within three hours of being contacted.

Depending on what's going on at the time with crew numbers across the airline's fleet, on some blank lines, you'll do a few standbys and one 'recency' flight to stay legally current. Other times, you can end up doing more flying than you would on a regular roster. To remain legally current and continue to fly, pilots must fly once every 45 days. This requirement varies in different countries. If they don't fly, for one reason or another, they need to fly with a check captain or do three take-offs and landings in a simulator to satisfy the licensing requirements.

A typical roster at the start of my career, which you bid for based on seniority, was 56 days of planned flying. If you were 'junior filth', you might end up going to Frankfurt in the winter, three times in a roster. If you were senior, you might spend a lot of time on the beach in Hawaii, Tahiti, or the west coast of the USA. The term 'junior filth' is a term of endearment among the pilot body, who are generally great people.

A good reason for knowing your rostering rules is to know your contact requirements while on rest days in an overseas destination. For example, you might arrive in Frankfurt, Germany, have three days off, and plan to drive into the Moselle River wine area to overnight and sample the fine teas available there. But, even though you aren't on standby, you still need to be contactable by the company in case some other pilot gets sick, or there's a change in your flight schedule. This means leaving a local landline contact number with scheduling or the local Qantas office.

The thinking guys have this all down to a fine art. The old favourite was, 'Oh, I'd love to go to work, but I just had a drink, so call me in eight hours.' Previously, the rules for alcohol and smoking were 'eight hours bottle to throttle and no smoking within 50 feet of the aircraft'. I used to love winding people up when they asked me what the rules were. I'd say, 'No smoking within eight hours and no drinking within 50 feet of the aircraft.' These days there isn't an hour limit for drinking. It's now a blood alcohol reading, and it's very low. There can be random drug and alcohol checks at any time at work, so pushing the limits would be very risky.

Another way to avoid getting contacted would be to be out when they called. They won't ring you on your mobile if you are out, as receiving calls on mobiles while overseas is expensive so you wouldn't answer anyway. Later, when they did manage to contact you, they would say, 'Second Officer Burfoot, we tried to contact you at 4 am but you were out?'

'Ah, yes. Please keep this to yourself, but I think someone may have spiked my tequila shots and I got tied up in an S&M club for a few hours.'

Or if you did happen to answer your room phone: 'Hello?'

'It's crew scheduling, would Second Officer Burfoot be available?'

'He's not here at the moment. Can I take a message?'

'Oh . . . we're looking for someone to operate the QF68 to Karachi in Pakistan, have a week off then fly to Kabul in Afghanistan. You'll have three days off there then fly an ISIS charter to Nigeria. Do you think he would be interested?'

'Definitely. I'll let him know.'

'By the way, who is this I'm talking to?'

'His wife.'

Pilots talk about financial stuff and what to do with their money a lot too. They all dream about retiring at 40 and sailing around the world on a yacht with a sailor, a cook and a masseur. To ensure you can retire at 40 with $100 million, you need to be on your game. Discussions on finance get interesting. I had one first officer do a hard sell on me about Bitcoin. He did a good job too, but the desperate nature of the sales pitch just confirmed for me that Bitcoin is a Ponzi scheme that relies on current owners pushing it to greater fools.

There was always someone pushing some scheme or another at other pilots, who are basically a captive audience. In the early part of my career, a couple of con men called Butler and Polenail started a scheme and got pilots to invest, promising large returns. Early adopter pilots encouraged other pilots to get in before they missed out. Some re-mortgaged their houses to get in further. Predictably, Butler and Polenail sailed off to the Bahamas or somewhere like that and everyone lost everything. It was easy enough to see who had been burnt in the aftermath, because those crew would only eat on the aircraft where it was free. They'd save their meal allowances while on the ground. The first thing they would do on boarding was find food, and the last thing they would do before disembarking was scoff more food. This became known as a Butler-Polenail diet.

Pilots are always very good at knowing where the cheap deals are on everything. If you wanted something like a mechanical bull for your party room, someone would know where to get a deal on one. 'There's a mechanical bull store just behind the sex shop in Blacktown, and if you mention Qantas, they'll install it for free.'

Pilots, in particular international pilots, find themselves with a lot of time on their hands due to the high-hour, long-range trips they do. To keep themselves out of trouble, many start businesses. These commonly become the topic of conversation on the flight deck in the middle of the night. From what I could see, this was because someone was either trying to sell you something, or they wanted some free advice on what they were doing.

When I think back to some of the businesses that were started or suggested, there was definitely some entertainment in these sessions. One guy was looking at starting an online burglar supplies shop with 'low vis' vests, noise reducing headphones (trust me, they would sell) and a loot bag of other ideas. I suggested that a business like that would be recession proof, as the worse things got, the more people would get into the burglary business. The other second officer, who was standing behind us waiting to swap seats after a break, suggested including get-out-of-jail-free cards for orders over a certain dollar amount. Pilots are nothing if not creative.

Other things? Well, we discussed HAFE, or high-altitude flatus expulsion, earlier. It may surprise you to learn that pilots aren't immune, so every now and then an indiscretion would occur, which would raise issues. There are many instances of this, but I have two favourites.

The first was when I was flying with Japan Airlines, and I had an American co-pilot and an American flight engineer. These two Americans hated each other as a result of some strike breaking that had happened during the demise of Eastern Airlines, from memory. Anyway, I had half-heartedly been trying not to fart, but in the end had given up and let it out. The cockpit is a rather confined space, which tends to concentrate things, if you know what I mean. Just at the right moment, I turned to these two with a look of disgust and said, 'Which one of you pigs did that?' Well, of course, each knowing that they weren't responsible, both assumed that it had been the other one. Then I said, 'Do you know what really disappoints me? It's that neither of you is man enough to own up and apologise.' They nearly came to blows.

On another occasion I was flying with an Australian first officer and he looked at me and said, 'Sorry, boss. That was me.' I immediately thought that this would be a good time to sneakily release one of my own, so I did. About 20 seconds later he looked at me and said, 'That's not mine!'

Pilots also love a good laugh, so the opportunity for a good wind-up is never missed. On one occasion back in the day, the US air force long-distance, heavy-lifting aircraft all had the callsign 'MAC', short for Military Airlift Command. One day, around the time the Gulf War was kicking off, we were in the area and there were a lot of MAC callsigns on the radio. The second officer noticed this and asked, 'Who are these Mac guys?' I told him they were McDonald's transport aircraft delivering fries and burger buns to all the Mickey Ds in the area. Sadly, he believed me.

On another occasion, we were over Afghanistan on a summer's day, and I pointed out a lake to the second officer. I said, 'That's where the Taliban go waterskiing and that line of mountains over there has some great ski fields on it in the winter.' He looked at me for a bit, finally saw a slight twitching in my cheek, and knew I was winding him up. As you can see, we pilots rarely allow things to get boring.

Pilots do like to look out the window a lot. In the daytime, some of the places we fly over are unforgettable. Seeing things from the air gives one a whole new perspective on things. Afghanistan is a classic example. The mountains are very impressive but there's not much green anywhere. It's like a mountainous desert that is covered in snow in the winter. And when you do see villages, you can see that they are all basic but secure.

You'll be pleased to know that knowledge of the systems and how they work also gets a run for discussion at some point during a flight. Modern aircraft are highly complicated and it's not necessary to know everything about them. But it can be fascinating, so quite often the two pilots will discuss and research how the systems work in the background. Junior pilots are generally like sponges and always keen to hear from more experienced pilots on anything useful.

This technical type of discussion is also quite common when one of the pilots is due in the simulator for a six-monthly check. It's particularly good to run the scenarios past other pilots and get their opinion on how they would handle it. It's also very useful running practice emergency drills in the cockpit and pointing to switches and saying what you would do. In this way everyone gets some revision.

Looking out the window at night can be quite amazing at times too. In higher latitudes it's not uncommon to see the northern and southern lights. Being away from other light sources and up high means the stars and planets are brighter, and you can see more of them too. As an ex-air force navigator who used to navigate the by the stars, I'm fairly knowledgeable. Back that up with an inter-active astronomy app and I become a genius. Just ask me.

In 1986 I was lucky enough to get views of Halley's Comet for a few months. That's significant because the comet has now gone on a long journey of exploration and self-discovery and won't be back until 2061, just after I turn 102. I hope my eyesight is still okay then.

Another thing that is always impressive is meteor showers, where a number of meteors, formed by meteoroids (or high-speed space detritus), show up at the same point in the night sky. The showers recur at approximately the same dates each year and vary in the speed, frequency and brightness of the meteors. Currently, there are 112 established meteor showers. Watching shooting stars all night is a great way to stay awake. There's not a lot of worry about one of these meteoroids hitting an aircraft. Most are too small and burn up in the atmosphere. If one ever does make it through to aircraft flight levels, it could, in theory hit an aircraft, but what are the chances?

So, looking out the window is definitely one of the perks of the job, which leads me to mentioning my fifth form maths teacher, one Dave Mill – a very nice man (RIP). He would stand up the front and write stuff on the board while talking at us in his gruff voice. I would start staring out the window, thinking about anything but maths. One time he turned around and said to me, 'Stop staring out the window, Burfoot. No one will ever pay you to stare out the window.' Well, I showed him, didn't I.

PILOT TRAINING

Training is central to a pilot's working life, and pilots accept that their career requires constant training and assessment. Training to fly a new aircraft, training for a command and the requirement to be checked every six months are examples of what every pilot must face. On average, a pilot can expect to face four 'recurrent' simulator sessions every year, plus a line check.

No other industry has this training and checking requirement. The training is demanding, and pilots, in general, do not like being checked.

Recurrent sessions are the mandatory simulator training and checking a pilot must undergo, two every six months, to allow them to exercise the privileges of a licence. The line check happens once a year, and on that, you fly a normal sector from A to B in the aircraft while a check captain watches on the flight deck. At the end of the flight, if you haven't scared him too much, you are signed off on the line check for the next year.

Part of the line check includes a 'question time', where the check captain discusses a few things that have been in the news lately and asks a few questions to ensure your brain has retained enough information to do the job. An example of a question might be, 'What are the radio procedures for unattended aerodromes [aerodromes with no control tower and radar]?'

If you don't know the correct answer, you could say, 'I can't believe you don't know that, Bill, what with you being a check captain and all.' Whereupon Bill says, 'I do know it, Jeremy, but I'm trying to establish if you know it.' At this point you change the subject by asking Bill a pre-planned question and see if you can get away with it.

So, pilots accept that training goes with the territory, but most pilots feel anxiety before a training event. Anxiety is often overstated, but deep and effective preparation is the key to managing these anxious moments. Some of the old-school trainers and checkers were feared so much by trainees that the trainees would often take sickies if they saw a particular instructor was rostered to run the session. Creating an atmosphere that promotes anxiety is totally counterproductive and may result in the trainee underperforming and not reflecting their actual ability. Knowledge and skills can be tested without belittling, sarcasm or continual faultfinding. It is very easy for anyone to say 'what' someone did wrong. It's far harder to analyse and diagnose the 'why'.

Fortunately, the training environment has improved over the years, and most instructors these days are very good at creating a learning environment before a checking environment. Despite this, flight simulators are commonly described as 'lurching caves' or 'the Tardis' by pilots. To get to the Qantas simulator building, you had to park your car then walk over a 100-metre overpass. This was known as 'the walk of death' and known to be similar in anxiety levels to the last steps of a death row inmate as he approached the electric chair. Some pilots allowed 20 minutes to cover the 100 m.

Modern flight simulators cost around US$60 million. They look like shrunken versions of the alien machines from the *War of the Worlds* movie. They are state of the art with full motion and visuals that feel and look like the real world. They are better than the aircraft for training, because they match the aircraft in performance and feel, but are cheaper to run and allow for practising all manner of emergencies with no jeopardy. The experience in them is so real that a bad weather session can leave you slightly depressed until you walk outside and realise that the sun is shining. A four-hour session in a simulator will also leave your brain so exhausted, on occasion, that you are incapable of anything other than finding the nearest pub and having a beer.

In my early years at Qantas, I was a simulator instructor as a first officer. My job was to train second officers and first officers. Qantas has used 'training first officers', or TFOs, on and off over the years as a cost-saving measure and to allow first officers to gain training experience before later becoming training captains. I thought it couldn't hurt to become a TFO for a few years and was ready for a break from flying full rosters. I'd still go flying once a month to keep current, but the rest of the time, I ran simulator sessions.

The training to become a TFO involved learning to operate the simulators and giving briefings. It also involved learning about instructional techniques and doing assessments. This mostly went pretty well but sometimes I wondered if the trainers training me to be a trainer had had any trainer training themselves.

After a few months of training for me, I was cleared to start instructing on my own. A training session would involve an hour of briefing and then four hours of flying and procedures in the simulator. I enjoyed it a lot, as I was learning incredibly quickly. You have to be on the ball and have all the answers, and it focuses you. You also get to see all the right ways and the wrong ways of doing things.

I continued as an instructor through 1988 and most of 1989. I could see that it was likely I could get a shot at an upgrade to 747 captain in 1990, so I was making the most of my time in the simulator. After evening sessions, if the simulator was free, I'd debrief my students, then get back in and fly that thing until I was worn out, sometimes until the early hours. I'd operate the whole thing by myself and did countless emergency procedures for practice.

Sometimes my students would come back into the simulator with me for hours, and I'd show them stuff they had never seen, then let them try it for themselves. By the end of 1989, there wasn't much I couldn't do in the simulator. My 'party trick' that I used to impress my students was to go under the Sydney Harbour Bridge towards the west at about 350 kt, pull up to 2,500 ft, shut off all four engines, then make a glide approach onto runway 16 at Sydney. If only I could have done that for real, but I suspect if a pilot did, these days, they would shortly be 'helping police with their enquiries'.

In the good old days before simulators were approved for landing practice instead of using the aircraft, pilots would have to do what was known as base training. This was required for upgrades to first officer and captain, and with Qantas, it involved flying an empty 747 down to Avalon in Victoria and zooming around the circuit there doing multiple landings. If it wasn't so stressful from the checking perspective, it would have been fun. The low-level circuits at 500 ft were particularly impressive. When doing turns at 500 ft in a 747, it seems like the wing is scraping along the ground.

When I started working for Japan Airlines in the early 1990s, on a leave of absence from Qantas, I had to go through base training all over again at Moses Lake in Washington State, USA. My training buddy was an American, George Palfi ('Big George', on account of the fact that he was about 5 ft 20 in tall and weighed around 130 kg).

Moses Lake is uninspiring, and being mid-February 1992, it was cold and dreary. They stuck us in some tragic motel on the edge of town. It was pretty depressing, really, but the flying made up for it. JAL had a crew room at the airport and an operations centre for briefings and the like, so the operational set-up was good.

The airport itself used to be a military facility, and it has a runway 4,110 m or 13,500 ft long. We met our instructor, Captain Honzawa, who seemed like a good dude. He gave us the complete briefing on a whiteboard. This was funny, because when he described the circuit we'd be flying, he drew all sorts of odd things on it, like the Target store, which we could use to line up on. When he drew the downwind line of the circuit, we expected a straight line, but he put a hernia-like bubble extension out to the right. Then he drew a house where the line should have been. 'This is Mr Bator's house,' said Honzawa. Then he paused and sucked a huge breath through his teeth. 'Most severe. You must not fly over Mr Bator's house,' he said. Mr Bator had apparently shot at a JAL 747 flying low level over his house in the past, and since then, the circuit had been crooked. Someone should have told Bator the war was over. That would have simplified things.

For the first four days, we flew with Honzawa-san. We'd climb up to about 10,000 ft, and then we'd practise steep turns at 45 degrees angle of bank. I can tell you that it's easier in the aircraft than in the simulator. With a bit of concentration, you can actually go from left bank 45 degrees through level to right bank 45 degrees without climbing or descending. After steep turns, we'd do full stalls and stall recovery, which, in a big aircraft like that, is quite something. For those of you who don't know, a full stall is where you pull the nose of the aircraft up so high that lift over the wing breaks down and the aircraft goes into a sort of unstable freefall. The recovery is to push the nose down, pick up speed and re-create lift over the wing. On the third day, we did all this in moderate to severe icing conditions in the cloud. This was just plain silly. I was not happy with that at all, but we survived.

After upper air work, we'd head back and fly an offset NDB (non-directional beacon) approach, then hit the circuit for standard and 500 ft circuits. These circuits were all good from a safety aspect, except that there was nearly always two 747s in the circuit. Quite often, we'd be making an approach to land with the other bird doing the same from the opposite direction. One 747 would always have to peel off at the last moment. It crossed my mind that the risk of doing this was right up there with doing upper air work in icing conditions.

Day five was our check day, with Honzawa-san in the co-pilot seat and a JCAB (Japan Civil Aviation Bureau) checker watching from behind us. The morning of the check, we got picked up early at sunrise. Big George said to me, 'Ahhh, Jer [sounds like Jair], it's a great day for flying, the sun is rising in the east, and Mr Bator is cleaning his guns.'

We arrived at the training centre and were marched in to stand at attention in front of the JCAB examiner. We had figured out that the deal with the Japanese was not to keep getting all the questions right. You needed to get two or three right and then get one wrong. Otherwise, they'd keep asking until you got one wrong. Being clever was not clever.

Big George and I had discussed this and confirmed our plan in advance. But it all fell in a heap from the beginning, because the JCAB bloke's English was so bad that we had no idea what he was asking. As a result, we got the first three questions wrong. I can still remember one of the questions. He said, 'How do the lights change?' What would you answer? We had no idea, so I said, 'With a switch.' Later in the debrief, he showed us a coloured picture of the runway lights changing from green to orange then red as the end of the runway came up.

After the supposed failure of the oral exam, we were marched out to the aircraft and launched to run through the routine we'd been practising for the last few days. One of the standard things on short final for landing was for the engineer to call 50 ft and 30 ft as you entered the flare for the landing. In my previous landings, I was having a great patch of super smooth landings, then on the final one, someone told the engineer not to call the altitudes. Luckily, I picked it up from my peripheral vision, flared a little late but slightly quicker and did a super smooth landing again, thinking, 'Fuckers. I showed you.'

A little while later, we were in the office again, standing at attention in front of JCAB-san. He paused for a minute and then said, 'You pass.' Then we rode back with JCAB-san via the grog shop. He bought some beer and said, 'You come to my room and drink beer.' This was an order, not an invite, so off we went. He offered us a beer each and cracked one himself.

Then Big George subtly pointed out his can to me, and I saw it was zero alcohol beer. JCAB-san had not seen this when he purchased it. After the second beer, George and I were wondering how we were going to get out of there, because JCAB-san was starting to act drunk. He was laughing a lot and patting us on the back. But just when we thought it would never end, he abruptly said, 'Okay. You go now.' Then the actual drinking started.

So training is definitely a huge part of a pilot's career. Techniques do not sit still, which means pilots need to keep moving to be across advances in aviation technology.

Pilots acquire up to 90 per cent of information visually, and a pilot must use a scanning method, which is moving the eyes in a structured and coordinated manner over the flight instruments, to get the necessary information to fly the aircraft. The scan that a pilot must learn is a fundamental skill, particularly as aircraft flight decks are becoming more advanced, with 'glass' flight instruments becoming the most common instrument presentation.

More and more information is being presented to the pilot, and this 'visually dense' environment suggests training methods will need to evolve. Eye-tracking technology has now advanced to make this technology attractive to aviation to assist with training. Eye trackers have the ability to show a flight instructor where a student is looking to help in training. Eye trackers are now starting to be fitted to some flight simulators and may well become a standard feature.

Virtual reality has also been used in flight training, but these devices will likely be expanded in how they can be used. For example, pilots could use virtual reality devices at home, perhaps as a mini simulator, where the pilots can practise eye scans. The scans may help maintain a higher proficiency by keeping scanning performance at much more accurate levels than before.

PILOT UPGRADES AND AIRCRAFT TYPE CHANGES

Probably the hardest thing pilots do in their career is to be upgraded from first officer to captain. Junior pilots are working towards this from the time they start in the airline and are always looking to learn from experiences.

I started with Qantas as a second officer. It's a good way to learn by watching and participating when the opportunity comes up. The downside to this was that you didn't get to fly the aircraft during take-off and landing. After a while your manipulative flying skills dropped off to a certain extent.

When my first officer upgrade started, I had to remember how to fly again. The jump from flying light commercial aircraft in the jungles of Papua New Guinea to flying a 747 was huge, and I struggled to get my head around it. My final check was a four-sector regional trip. On approach into Bali, I was still coming to grips with the sheer size of the 747, and I let it get the better of me. The aircraft was controlling me rather than the other way around. As a result, I got a little bit high on the final approach and the check captain made me reject the landing and go around for another go.

This is a big deal in airline operations, so I failed that sector. On the following sector my confidence was shattered, and I wasn't flying well so I failed overall for the check. This resulted in a bit more training and then another final four-sector check to London and back.

Because I'd already failed once, there was even more pressure on me for this check, so it was not an enjoyable time. Fortunately, I was able to impress the checker enough to be passed and become a first officer.

Even though there was a rapid expansion in Qantas, my captain training came up way before it should have. There were a couple of reasons for this. The first was that the experience levels of the guys around my seniority level was low. The command course was known to be very difficult, and you only got two goes at it in a career, so many of the guys chose to put off their training for a while. The second reason was that quite a few guys around my vintage wanted to fly the new 767s.

At the end of February 1990, I started the course. I thought I was mostly ready. I certainly had all the systems and simulator flying sorted. The only problem I could see for my command training would be the route flying. I had minimal experience with that, especially since I had spent the last couple of years mainly on the ground in the simulator. Therefore, I hoped that my aircraft handling skills would allow me to apply most of my brain space to the route operations and compensate for my lack of experience.

Command training in any decent airline is hard. It needs to be. You're going to be responsible for a $400 million aircraft and over 400 people's lives. That's a big deal. You can't screw it up. Accidents can cause irreparable damage to an airline's reputation, not to mention the financial damages. Some accidents have caused airlines to go out of business. So, before they sign you off as a captain, they need to be sure that you will cope and make the right decisions, even under the worst stress and circumstances, whether that's bad weather, fatigue or other external influences. The whole process is about piling stress onto you and seeing how you perform.

I worked extremely hard throughout the four-month course. I stopped drinking alcohol for the whole time . . . No, I did. Really. I didn't want to leave anything to chance. The simulator part of the course was relatively straightforward after being a TFO, but there was still the final simulator check to go. In those days, that was a simulated flight from London to Frankfurt in the dark and in the middle of winter with crap weather, icing and fog. I knew how it would go to a certain extent, from other guys who had been through it. At some point on the final approach to Frankfurt, something would happen to make you divert to somewhere else in Europe, like Paris or Amsterdam. Typically, right after you started diverting, an engine would fail, among other things, and so the rest of the flight was flown on three engines. It was a workout.

Because this check was a big deal, the company decided it wouldn't be fair if they called in a first officer who hadn't had time to prepare for it, so in those days, you could dial a friend. I rang a mate called Butch who was fairly experienced and asked him to be my first officer. He was flattered, of course, but also wary as he had his command training shortly after mine. He said, 'What if I screw it up and look like a dick?' I said that this was highly likely (joking) and reminded him of the value of seeing this final check before having to fly his own one. Butch saw the logic and agreed to do it.

We turned up on the day and had an hour of questions, then jumped in the simulator. Soon we were airborne and heading to Frankfurt with everything under control. Once we diverted from Frankfurt and all hell had broken loose, I demonstrated my delegation skills and watched Butch working like a one-legged man in an arse-kicking contest. He remembers coming up for breath at one point and seeing both the checker and me sitting there with grins on our faces watching him sweat. I allowed him 0.2 of a second to take a breath and then gave him some more things to do. In the end, we survived, and that box was ticked off.

The base training in the actual aircraft at Avalon was also fairly straightforward, and I actually enjoyed flying around at a low level again. There was a final check for that, too, which went well.

The final phase of command training was route flying. In this phase, you had to show excellent decision-making, good route knowledge and sound risk management. As well, your leadership had to be up to scratch so as to get the best out of your crew. You also needed to demonstrate the ability to keep the bigger picture in mind rather than focusing on things that weren't important. And you had to run the whole operation efficiently.

I went on a few training trips with various training captains and, after about six weeks, was put up for my pre-final command check. This was a check to see if I was ready for the final check. I did this with Captain Col Mercer, who was one of the all-time good guys. On all my previous training trips, I'd been doing great landings every time, but for some reason, my touch had deserted me with Col. Fortunately, the landings weren't too bad, but every time I'd do one, Col would laugh. In the end, he was happy to pass me, however, so that was all good.

A week later, I headed off with Captain Rod Engledow to Singapore, Bahrain and Frankfurt on my final check. On the ground in Bahrain, prior to departure, the engineer reported a fuel leak from one of the engines. This was stressful as it was near the limit for what was allowable and took some time to sort out. Yes, at that time, a certain amount of leaking fuel was legal, but we are only talking about a drip here, not a flow. In the end, we departed and arrived in Frankfurt with no further drama.

I now had one sector left to get checked out. That happened a few days later. We flew from Frankfurt and landed in Singapore on 27 June 1990. Then we taxied off the runway, and Rod said, 'Stop here.' The second officer leaned over and replaced my three stripes with four stripe epaulettes, and Rod said, 'Congratulations, Captain Burfoot.' Six years and one month after joining Qantas, and at the age of 31, I was a 747 Captain. Some serious beer was drunk that night. I was thrilled. The only thing I regretted was that Dad hadn't lived to see that, having passed from cancer in late 1989. He would have been very proud.

Because the upgrade process is so intense, many pilots never make the grade. After two unsuccessful attempts, they are consigned to being a permanent first officer. There can be any number of reasons why. It could be from lack of hard work or an inability to focus on what is important. An example of this was a first officer I flew with many times. He was a lovely guy but just never seemed to get it. He would continually focus on detail that wasn't relevant. Prior to one arrival into Taipei, Taiwan, he spent 30 minutes briefing myself and the other pilots. I'm pretty sure he included his intentions if China invaded while we were on final approach. At the end of the brief, he looked at me and asked how his briefing was. I had to tell him that after ten minutes had passed, everyone had nodded off and besides that, he had briefed for the wrong runway.

Some people have it, and some don't. It doesn't mean they can't make excellent first officers. I don't know of anyone who left the job rather than become a permanent first officer. Why would you, with the lifestyle and the pay? Once your damaged pride has healed, there's a lot to like about being a first officer.

A good characteristic for a pilot is to know your limitations. If you know you are not the sharpest tool in the toolbox, but you work hard, that's okay. The pilots I didn't like were the ones who thought they were good, but they weren't. That's dangerous, and annoying to watch.

The second hardest thing for a pilot to do is to change aircraft type. When I first started my A380 conversion course in 2017, I was pretty apprehensive. I hadn't done a conversion course since the 747-400 in 2002, which was just an upgrade to the 747. With the Airbus A380, it was new technology and a whole different way of doing things. I'd also heard stories of old farts who had preceded me and struggled with the course. The general thinking was that they had tried to bring their Boeing procedures with them, rather than accepting the new system and getting on with learning it. In other words, they were still trying to tango at a disco.

I also wondered whether I still had the capacity to spend that much time learning again without nodding off every 20 minutes. It turns out I did. Thankfully, I got excited about the technology and the sheer size of the aircraft, and my enthusiasm put me on the crest of a wave, which I rode right through to the end.

From a specifications point of view, there were quite a few differences between the 747-400 and the A380-800. I've listed some below:

	747-400ER	A380
Max Take-Off Weight	410 t	575 t
Max Range	14,200 km	14,800 km+
Fuel Capacity	241,000 L	320,000 L
Typical Passengers	410	555
Engine Thrust	c. 62,000 lb × 4	c. 80,000 lb × 4
Wing Span	64 m	80 m
Main Wheels	16 wheels @ 49 inches	20 wheels @ 55 inches tall

Everything else isn't much different, except that the A380's upper deck extends all the way back.

From a pilot's perspective, there are some significant differences. Instead of a control column between your legs, there is a joystick on your left, if you are the captain, and on your right, if you are the first officer. The control column has been replaced by a pull-out computer keyboard that converts into a dining table. These minor differences can take some getting used to, especially after a long period of using another technique.

The thing other pilots warned me about was the thrust levers. They said they would screw with my head because, with the automatic throttle engaged, they didn't move up and down with the thrust like they do on a Boeing. On an Airbus you manually set them at the upper limit of the thrust range for the phase of flight and leave them there. Once you have set them at that upper limit, they don't move, even as the thrust goes up and down. The only indication you get of changing thrust is from looking at the engine gauges or the engine noise level. It's a weird concept, but brilliant, and I actually loved it.

As well as this, the A380 is fly by wire, which replaces the conventional manual flight controls of an aircraft with an electronic flight control interface. The movements of flight controls are converted to electronic signals and then transmitted by wires. The latest fly-by-wire systems will figure out which control surface positions are needed to achieve a pilot's plan. The system controls the rudder, elevator, aileron, flaps and engine for each situation, meaning the pilot can rely on it to react as expected. In this way, the fly-by-wire systems ensure pilots remain within an aircraft's safe performance range.

Apart from these differences, the change in size wasn't that noticeable. I imagine coming from an A320 might be a big jump, but apart from the wingspan, there wasn't a huge difference from the 747. In fact, it was easier to taxi around, because cameras and indicators would show you on a screen where the wheels were sitting relative to the taxiway centrelines.

The engines were a lot bigger, but so was the weight, so performance wasn't much different, except that the massive wing meant being able to climb earlier to higher flight levels.

All in all, the Airbus system is a pretty good system, and contrary to what I had always expected, I liked it. Moreover, I took to it reasonably well once I learned all the new calls and 'Airbus speak'. And so, on 30 November 2017, I flew my first flight in command on the A380 from Sydney to Dallas, Texas.

PILOT CREW REST

By now you must be exhausted from reading about all the hard work involved with pilot training, and I'm certainly exhausted from telling you about it, so it's time for me to visit the A380 pilot crew rest. You can't come, because that's banned for security reasons, but I can tell you about it.

Aircraft crew rests have varied over the years, from spare business class seats to bunk set-ups behind curtains on the flight deck to the ultimate crew rests on the A380, where each bed is in its own separate cabin and the sharing of high-altitude flatus expulsion is not mandatory.

The A380 crew rest is about as good as it gets. I could tell you that the captain's crew rest has a spa, a personal flight attendant, personal chef, a massage table, wine fridge, an outside sundeck and a 100-inch QLED TV linked to a PlayStation 5. But I won't, because that would be fibbing. We were led to believe this would be the case when Airbus was running around selling the concept of the A380 to airlines. The A380 would be like an ocean liner in the sky, they said, with cinemas, petting zoos, strip malls, and the chance of holding the Super Bowl on one was said to be highly likely one day.

But once the aircraft were being built and the accountants stuck their noses in, reality hit home, and soon they were looking at employing Japanese *oshiya* (train pushers) to push the last few passengers on to the aircraft. As a result, the captains crew rest was reduced to a room about 2 metres long by 1.5 metres wide with a bunk, a chair and a 7-inch LCD TV.

This isn't too bad at all, and it's a nice quiet place to take a nap.

Most pilots of long-haul flights operate the roster on flexitime, whereby the length of the time off is based on need. Everyone gets the same amount of time off, but you can take it when and how you want, as long as you work in with the other pilots. It's a great system. A certain amount of rest is required when operating a long-range flight. This helps ensure that pilots are in good shape for the landing despite having been on a flight for up to 18 hours.

Crew fatigue is a big deal. Airline companies claim to monitor it, but they are not really serious about it. In the end it's up to individual crew members to decide when they are too tired to work, and then to take sick leave.

I once got rostered for a trip that involved a three-sector tour of duty overnight from Singapore to Bali to Melbourne to Sydney. The duty time was close to 14 hours, so they gave us a second officer. The problem was that the legs were so short, none of us had time to have a decent sleep. On the way into Melbourne, we busted an airspace limit. It wasn't unsafe but it was still report-able, so I had to phone the duty air traffic controller after we landed in Melbourne. He was less than impressed, of course, but I mentioned we were all tired and apologised. He said, 'All good. Go home and get some sleep.' I didn't mention that we still had to fly to Sydney. After that experience, I refused to do that duty again, even though I got rostered for it. I just called in sick. Eventually everyone was avoiding it, so they changed it.

Long-term fatigue is insidious, because you get used to it and it feels normal. Then you go on leave, and after a few weeks, you wonder what the unusual feeling of health and vitality is all about.

One thing that's become evident to me after many years of flying is the toll that the job takes on your health. The jet lag, the dry environment and the irregular and reduced sleep all can't be good for you. My eyes used to have a yellow tinge to them, but now that I have stopped flying, they are white again. I'd come back from each trip with swollen ankles from fluid retention. That doesn't happen anymore either. I now realise how stressful the job can be with medical checks and simulator checks, and operating at the highest standard every time you go to work, no matter how tired you are.

Fatigue is a funny thing, though. I've had times when I've crossed so many time zones over a number of weeks that my body had no idea where it was. It's not a feeling I recommend. Sometimes you're not sure if you are tired or not. Other times you'd be struggling to stay awake sitting in the seat, but when you'd go to your bunk for a rest, you couldn't sleep.

Controlled crew rest is approved in many airlines, especially with two-man crews and no crew rest. This involves one pilot napping in the seat while the other monitors. This is probably better from a safety perspective than both pilots trying to battle through and stay awake all night. It's a bit risky, though, because if the other pilot nods off, no one is watching the ship. This is why modern aircraft have alarms that go off if no one pushes a button for a certain amount of time.

Having a pilot napping when cabin crew come up is not likely to inspire confidence, so one enterprising pilot with an engineering background suggested an easy design to get around this. It involved a metal tube attached to the top of the control column with a chin holder on the top of it. Pilots could be asleep, and the chin holder would hold the head up straight so the pilot would look awake from behind. The chin holder would also have a small cup in it to collect drool. These should be seen on flight decks shortly and are probably just stuck in a container on a wharf somewhere.

And now airlines are gearing up for flying non-stop halfway around the world. This, of course, is the absolute limit, because why would you bother going more than halfway around the world? It would be shorter to go the other way. Despite this, we are still talking about flight duty times for pilots of up to 23 hours. Bugger that. I'll stick to writing, thank you very much.

WHAT CABIN CREW
DO IN THE CRUISE

We talked about cabin crew earlier in the book briefly, but now it's time to discuss them in greater depth. I have mentioned my admiration for them. I couldn't do that job, mainly because I would biff someone for being rude on the first day. The job does take a truckload of patience and tolerance. Maybe that's why my marriage lasted as long as it did; Manola definitely needed the truckload of patience and tolerance for it.

Occasionally, even cabin crew can fail at this. In 2023, an American Airlines flight from New York to Georgetown, Guyana, turned back after two hours due to a 'disruptive passenger'. A business class passenger had asked for assistance in putting a bag in an overhead locker due to him having had recent spinal surgery. The flight attendant replied that he didn't get paid enough to do that. Later, the passenger referred to that flight attendant as 'waiter' in a harmless way. This caused the flight attendant to threaten to turn the aircraft around and return to New York. The passenger challenged him to do that. The flight attendant ran to the flight deck and convinced the captain to return to New York.

The passenger later got a full apology from the airline. If that was my airline, I would have fired both the flight attendant and the captain.

Given the sheer number of cabin crew, a small percentage will always be outliers or dickheads, just as in the pilot ranks. My intention is not to focus on these special cases any more than the others, but some of the stories about them are gold.

It was quite common in the early years of my career to hear of cabin crew sessions in the lower lobe galley or other secluded areas where alcohol would be consumed. In general, it was an open secret, and no one seemed too fazed about it. A pilot friend of mine caught them at it one time and had to shut it down. But he didn't report it. I'm sure it still goes on these days, but no one boasts about it anymore.

As with pilot nicknames, the cabin crew also have their list of names bestowed on those cabin crew who earned them. Here are some of them:

Wicket Keeper – puts on gloves and stands back
Harvey Norman – three years, no interest
Sensor Light – only works if someone walks past
Noodles – thinks all jobs take two minutes
Blister – appears when the hard work is done
Show-bag – full of shit
Lantern – not very bright and has to be carried
Deck Chair – always folds under pressure
Perth – three hours behind everyone else
G-spot – you can never find him
Bushranger – holds everyone up
Wheelbarrow – only works when he's pushed

Limo – carries about eight people

Cordless – charges all night but only works for two hours

As passengers, you probably have a better idea of the sorts of jobs cabin crew have to do inflight than I do, but I want the cabin crew to get full credit, so to make sure I covered it all, I asked an old cabin crew friend for a list. The 'Baroness' and I flew together often, so she knew my coffee preference and she called me 'Captain Long Black'.

When I asked her to send me a list of what cabin crew do, she sent me this:

Sure, list as follows:
1. Make coffee for pilots
2. Feed pilots
3. Put pilot's beers on ice for post flight
4. Listen to boring pilots and pretend to be interested
5. Clean pilot toilets
6. Bow to pilots
7. Shoeshine for pilots (was going to say polish their shoes but you may have taken that the wrong way)

List is by no means exhaustive.

So, the Baroness has pretty much nailed it, but there are other less important tasks that they do as well. Let's explore these further.

Food service. Of course they do. This is obvious and needs no further explanation.

Clean toilets. While it's accepted that the best way to clean an aircraft toilet after six hours of use would be to hose it down with sulphuric acid and wash it out, the run-off from this presents a problem. Neither is it okay to spray petrol in them and set fire to it, as fire on aircraft is discouraged. So, cabin crew do have the job of donning gloves and going around to spruce up the offering. This involves cleaning up spillage and using a cloth to remix the germs on the benches. For the germs, this is not an actual threat, it's more like changing partners in a line dance.

Water runs. On full-service airlines, cabin crew walk through the cabin from time to time and offer water to passengers. This is a nice touch and very welcome, and may go some way to reducing medical emergencies and the associated paperwork.

Medical and emergencies. Any time the shit hits the fan onboard, it's the cabin crew's job to deal with it and keep the pilots informed. On many airlines, the cabin crew get a certain amount of emergency medical training such as using defibrillators and cardiac massage techniques. They also know where all the emergency equipment is kept, from fire extinguishers and smoke hoods to oxygen bottle and masks. They are there to look after you if your luck runs out, so it pays to be nice to them. We'll cover emergencies later in the book.

Prepare to land the aircraft if the pilots both die. According to many movies, flight attendants are also back-up pilots who come to the rescue when both pilots fall out the window or simultaneously get beamed up to the starship *Enterprise* by mistake. Sadly, the reality is that unless they have some sort of pilot training, they wouldn't have a clue, and you need to trust me on this when I say, you don't want that happening when you are a passenger on a flight.

Taking care of animals. At times, passengers take dogs on aircraft for support. An example of this is a dog accompanying a blind passenger. These dogs are known as service animals.

The dogs require their own seat next to the passenger and must sit on pads. Despite this, there are often 'accidents' and it's the crew who have to deal with it. Genuine service animals are a must, though, so airlines need to cater well for them.

But inevitably, in this day and age, all sense of reasonableness on this has been 'cancelled'. Emotional support animals (ESA) have become the new big thing. ESAs are animal companions that give some form of benefit to an individual with some type of disability. ESAs are intended to provide 'companionship and support' that helps alleviate one or more aspects of the disability. So, you can see this describes Paris Hilton and her Chihuahua perfectly. And therein lies the problem. People thought it was okay to take their accessory pets on board in their handbags or on a leash.

This stupidity is not limited to Paris and her ilk, though. A New York artist, named Ventiko, attempted to fly with her ESA, a peacock named Dexter, in January 2018. According to Ventiko, having Dexter by her side has 'really changed her life in a positive way'. Dexter was eventually denied entry onto his flight. And in November of 2014, right around Thanksgiving, a woman and her emotional support turkey were permitted to fly on Delta Airlines, with the owner escorting the turkey in a wheelchair throughout the airport. There is no report on the fate of the turkey leading into Thanksgiving dinner.

Miniature horses were a common ESA carried on aircraft in the past. And last, but certainly not least, is the story about an emotional support duck named Daniel, whose entry onto a 2016 airplane flight went viral on Twitter. Sporting red shoes and a Captain America diaper, passengers couldn't help but share pictures of Daniel on social media. Reading this inspired me to want to start carrying a loaded shotgun with me on flights. All this silliness increased the workload for cabin crew too, of course.

Surprisingly, considering the general madness in the world, in 2020 the US Department of Transportation ruled that ESAs are no longer considered service animals. So many airlines have moved to ban ESAs that have not been certified as a service animal. One airline even went so far as to ban animals with tusks. How dare they?

Sometimes cabin crew just have to suck it up and go above and beyond. Here's a story from a female cabin crew friend of mine: 'Once on a departure on a 747-400 on the upper deck, I was seated and strapped in ready for take-off. Just as we were about to turn onto the runway, an overhead bin lid opened. The bin contained duty-free bottles! This would be a disaster if we left it as it was, so I quickly unstrapped and ran forward to close the bin lid as the aircraft started screaming down the runway. Once that was done, I was stuck. There was no way I could get back to my jump seat safely without being flung down the cabin. So, I stepped in front of a business class woman and crouched down pretty much right between her legs! Once it was safe for me to leave her, I said, "Well, that was a first for me." She replied with a wry smile, "Me too."'

And another: 'Bombay/Mumbai flights were notoriously eventful flights for the cabin crew, mostly hard work, really hard work! One time in the middle of the night on the 747-300, all was quiet in the economy cabin, thank heavens! One of the crew had just done a round of toilet checks and came back to the galley to tell us he had blocked off one of the toilets. Of course, we asked why (with a full load of passengers in economy, every toilet is needed).

His reply was that someone had done a number two in the hand-basin! We decided to keep it locked.

'Later, our flight service director (FSD) came down to see how we were all going, so we let him know about the blocked-off toilet. He asked who on earth did that in the hand basin. We had our suspicions and one of the crew pointed out four passengers in a centre row of seats. Political correctness prohibits me from explaining why the cabin crew suspected these four passengers. The flight attendant doing the pointing actually meant *one* of those four, but our FSD misheard and thought it was a particular one of the four.

'The FSD took off in a hurry and soon came back with rubber gloves and paper towel rolls. He went over to one of the four passengers and instructed him to go with him to the blocked toilet. The FSD then came back to the galley. We were horrified that he had made that poor guy clean the hand basin. We then asked the FSD, "How did you know it was that guy?" He turned white and said, "I thought you said it *was* that guy."'

And another: 'While on a flight from the USA, working in economy class, we were delivering the breakfast service. After I had placed a breakfast tray in front of an elderly male Texan passenger, he asked me in his heavy Texan drawl: "Excuse me Miss, could you please tell me, what is this OOT milk?" It was actually UHT, a brand of long-life milk made in Australia. I couldn't help myself, so I answered, "Well, sir, in Australia we have bred a special type of cow called an UHT (pronounced as OOT). This cow produces milk that has a long shelf life." To my surprise, he bought it and said to his wife, "Did you hear that, Dolores? A cow that produces long-life milk. Well, I never." My colleague working at the other end of our cart leant over and said to me, "You're evil." We both laughed and moved to the next row of seats.' My apologies to any Americans reading this, but we often had fun with the Americans.

And now it's time to mention a few of the shadier things that entrepreneurial cabin crew have done. One such case was a scam to do with spirit miniatures from first class. The alcohol was free in first class, and the chief steward would give numerous bottles to the senior stewards in economy. The first round of unopened bottles would be sold to economy class passengers for cash, and after recovering the spent miniatures, they were refilled from large duty-free bottles and sold again. The funnel used for this was a paper cup with a biro hole in the bottom. Out of courtesy, crew made a snap verbal noise to simulate opening a new cap when giving it to a customer. Over an 11-day London trip, the dollars made from this scam were quite substantial. Some ingenious crew also supplemented their stock with cheap duty-free purchases before flights. Free booze for all classes inflight ended this practice.

With the advent of in-flight entertainment, there was more money to be made by enterprising cabin crew as some sold headsets to passengers in economy class even after they became free.

When the 747-300 was mainly doing Sydney–Tokyo return flights prior to its retirement, these flights were crewed by very senior flight attendants, because the Japanese yen allowances were lucrative when converted to other currencies. The same crew members would normally fly to Tokyo on the same day of the week. It got to the point where the crews conspired and rostered one or two of themselves off on a rotating basis for the whole flight. Instead of working, they sat in business class and ate and drank like normal passengers.

CABIN CREW REST

The Tokyo flight scam was taking the crew rest concept to an extreme, but there is no doubt that on long flights, rest is definitely required for cabin crew as well as pilots. The one major difference between pilot crew rest and cabin crew rest is that with the large cabin crews, the rest area contains many beds in a communal bunk style set-up. It must be hard to get any really good sleep given the comings and goings of crew and the sharing of HAFE.

On long flights, when there is a large cabin crew of 20 or more, the crew rest area must be large enough for half of them to be off at a time. During the long night hours, there isn't much to do in the cabin, so they might as well be sleeping and doing their job better when they come back on duty. So, after the meal service and clean-up is complete, a roster of time off is drawn up and the first lot go to the bunks. They could all end up with two lots of three to four hours off on extra-long flights. A couple of hours before arrival, everyone comes back on duty to serve meals and prepare the aircraft for the arrival.

These communal crew rest set-ups are also, occasionally, a recipe for trouble, and as you can imagine, all sorts of things have happened in them. I personally remember when one such crew rest was in the ceiling adjacent to a galley. I was a first officer at the time, and in the middle of the night I did a walk around the cabin to check that everything was okay. We were on a 747SP, and when I approached the rear galley, the curtains were pulled, so I pulled them aside and went in to say hello to the flight attendants.

There were two guys in there, and one of them said, 'Good timing Jeremy. You're just in time for the lingerie show.' Moments later two of the girl flight attendants came out of the crew rest wearing nothing but see-through lingerie. My being there didn't seem to bother them, and we were given quite a show. Then they disappeared back into the crew rest. I walked back to the flight deck wondering if I could believe my eyes.

INTER-CREW RELATIONS

Relations between pilots and cabin crew have ranged over the years from excellent to marginal. The difference in pay scales and time off has often been a contentious point, and the answer to that is that we all have choices. You can throw your lot into a pilot career, make the sacrifices, or not. Indeed, some cabin crew have retrained as pilots and are doing well, and good on them.

That said, many highly skilled people such as doctors and lawyers become flight attendants to see the world and have a change of scene. Never underestimate the possibilities of what cabin crew were doing in a past life. And as I've mentioned already, I have met many fantastic people in the cabin crew game.

In general, working with cabin crew has been a whole lot of fun. Of course, many pilots marry cabin crew, and there are even more unsuccessful romances that happen short term. As mentioned, I met my wife, Manola, while flying.

Carrying out a relationship with another flying person is difficult, because often when you are away, they are home and vice versa. Sometimes you would pass your partner in an airport terminal, heading in completely different directions. In the early days there were no video calls; the only communications were by landline, which used to cost many dollars per minute, especially if you called from a hotel room. These long-distance, part-time relationships were intense and stressful, but they had their good points. It was well known that when you got home from a trip and your partner was there, the second bang would be your suitcase hitting the floor.

Once in a while, we were lucky enough to fly on the same trip. We had some fun with this. On one trip I didn't tell the pilots and engineer that my girlfriend was onboard. She kept popping up to the flight deck any chance she got, and the engineer mentioned that this was unusual. I convinced him that she was after him and he got quite excited. We all had a good laugh later when the truth came out.

A flight attendant friend of mine called Heidi tells of how she first met her current beau on a flight to Hong Kong: 'As a flight attendant, one of the delights of the job was being able to join the pilots on the flight deck for either a take-off or landing – or both! I had so many, I couldn't possibly count them all. One in particular was a standout from the rest. It was on a flight from Sydney to Hong Kong on a 767 landing at the old Kai Tak Airport. The old Hong Kong Airport landing was definitely the most spectacular landing you could witness. On these flights, out of Sydney, at the pre-boarding stage, when the cabin crew saw that the pilots had arrived on the flight deck, there was always a stampede to get to the phones to call the captain and ask for the landing on the spare jump seat.

'I was first on this particular occasion, and our lovely captain gave me the okay. Now all I had left to do was to ask the flight service director for permission to do so. It was granted! So, once we had begun our descent into Hong Kong, I completed all my safety checks, bolted to the flight deck and strapped in behind the captain who was doing the landing. I'd seen this amazing approach a few times before, but my hands were still sweating, and I had a massive adrenaline rush. After we landed and had turned on to the taxi way, I said to the captain, "That was just amazing." He turned around and said with a very cheeky grin, "Best fun you can have with your pants on." We met again in later years; he is now the love of my life!'

And then there were always the 'short term' relationships that occurred from time to time. A flight attendant friend of mine told me how she had partied with the captain and first officer once in a hotel room and the three of them ended up in bed together. The next day, on the flight to wherever they were going, it was suggested that the three could become famous if they locked the flight deck door and had another threesome on the flightdeck. So they did. Thank God for autopilots.

Apart from relationships, there has always been a lot of fun working with cabin crew. Here are a few examples.

One time a flight attendant called Miles called up from down the back and said, 'Hi, it's Miles down the back.' The first officer responded, 'Yeah well, we're miles up the front. What can we do for you?'

'I'm sorry, captain, but I missed your PA, and I was wondering if you could tell me what you said because I have to translate it into Italian.'

'Well. Nothing much really. ETA 0645, temperature 27 and nuclear war just broke out in Europe.'

'Thanks, captain.'

On another occasion I did an impromptu wind-up of one of the cabin crew that was worthy of a gold medal. They'd also give us a bit of a wind-up from time to time, so it was considered fair game. We were over Afghanistan in the daytime, and one of the girls came up for a look and started to talk about how much nicer the air was in the cockpit than in the cabin. It's actually the same, but I launched into a speech about how that was a result of the BART system or Bad Air Retraction Turbine. I told her each cockpit seat had a fan in it which sucked the pilots' farts (HAFE) into a tube and then pushed them into the main aircraft cabin at door right two. Door right two was where the cabin crew rest seats were, and where they sat for their meals, but this detail went unnoticed.

She was impressed, and I finished by saying, 'They were going to name it the Foul Air Retraction Turbine but thought better of it.' Then she leant forward, looked out the window, and saw the pollution of Kabul out to the left. She asked, 'Where is that?' The first officer dove in now and said, 'That's Kabul. I'm surprised you haven't heard of it yet, because the company are going to send all the cabin crew girls there for attitude adjustment courses with a mob called the . . . umm . . . starts with T . . .'

'Taliban?' I said.

'Yeah. That's them,' he said.

'Awesome,' she said. 'Do you know if the shopping is any good there?'

The cabin crew were good at the odd stitch-up themselves. Once when I was a second officer, my trousers went 'missing' from the hanger in the crew rest while I was asleep. When it came time to go back on duty, I had to do it in my underwear. I knew it was a stitch-up, but I didn't know that just about the whole crew was involved. Over the next few hours, we had visits from nearly all of them. Thankfully the trousers were returned before landing.

Another time I colluded with the upstairs flight attendant on a flight from Cairns to Nagoya during the day. I'd seen her in the supermarket in Cairns earlier that morning and had asked if she was shopping for ingredients to make us Devonshire teas on the flight. Her initial thought was to tell me to fuck off, so she did. Then we agreed to wind up the first officer, so I bought all the ingredients and smuggled them onboard.

Once things settled down in the cruise, up came Giselle, the flight attendant, to see if we wanted anything. It needs to be said right here for context that Giselle was half my age and extremely hot, so when I said, 'Some Devonshire teas with scones and cream would be nice, my darling,' the first officer nearly wet himself laughing. But when Giselle leant in towards me and said with a very sexy voice, 'For you, captain, I would do anything. Anything at all,' the first officer quickly stopped laughing. Off went Giselle on her mission. She returned 20 minutes later with the hot scones and cream and jam. It was delightful. I never let on to the first officer, either.

Heidi tells a good story about winding up the pilots from a cabin crew perspective: 'Pranks were rife among crew. On one flight from Frankfurt to Bangkok, I was particularly bored in the upper deck galley on a night sector. We had a young second officer. The young ones were generally quite naive and fair game! So, with the permission of our captain, I took the jacket and hat of the second officer and went to work in my galley with one of my many hotel sewing kits. I zig-zag pattern-stitched the underside of his hat, so he couldn't put it on. Then I sewed the lining of his jacket sleeves together and sewed one of our Qantas finger puppets, which was a pale blue uniformed pilot (giveaways for the kiddies on board), into his jacket pocket.

'Upon arrival in Bangkok, it was very hot as usual, and the captain instructed the pilots to wear hats and jackets off the aircraft. Once the passengers had departed, the flight deck door was opened. I hung back to watch the ensuing comedy. Sure enough, I could just see the second officer trying to get his jacket and hat on. Neither would work and the swearing began. I decided I needed to leave the scene of the crime. Later, at the baggage carousel, I sidled up to the second officer and told him how much I loved his finger puppet. My comment didn't go down well at all!'

A relatively high percentage of cabin crew are gay and lesbian, so it was inevitable that I got to spend a lot of time with them, and I developed a good understanding of them as well. They do have an extraordinarily fun outlook on life and tend to live more for the moment from what I have seen. Some are openly gay and flaunt it, and you would never know others unless they told you. The ones who flaunted it were always entertaining.

One time I was passengering on a 767 in business class during the day. I was near the back, and I could see the business class flight attendant serving people at the front, slowly making his way in my direction. His name was John, and I knew him from an Athens trip we had done years before. As John got to the seats in front of me, I could hear him talking in a strong French accent to the passengers. When he got to me, he didn't recognise me at first and started waffling with the French accent. I held my hand up and said, 'John. It's me, Jeremy. I know you're not French.'

He threw his head back and laughed and said, 'Ohhh. French accents are all the rage in Sydney's eastern suburbs right now.' My life is far richer from knowing people like John.

Tragically, I joined the company in 1984 just as the AIDS crisis was taking off. It was terrible. You'd fly with a beautiful human being on a trip and then never see them again. Eventually you'd hear in conversation that they had gone.

DIFFERENT CREW CULTURES

Having worked with Japan Airlines for four years, I feel I can comment on the difference in crew cultures at least between Australian, Canadian, American and the Japanese, as our basing in Anchorage included crew from all these nations. Of course, we were working for the Japanese, so we got good exposure to all of their foibles. There was always a minor translation problem, which added humour from time to time.

On one occasion, I boarded the aircraft in Tokyo, jumped into my cockpit seat too quickly and ripped my trousers. Later, when one of the Japanese girls came up, I asked her if there were any sewing kits on board. She looked worried and said, 'I have a look, Captain.' Then she disappeared. Twenty minutes later, she was back, looking even more anxious. She said, 'You know that thing you wanted, Captain? They all been eaten already.'

The Japanese pilots were different too. They were a pleasant bunch and treated us well, despite the fact that we were allowing the airline to undercut their wages. But they were definitely different. For example, if someone dropped an industrial strength fart on the flight deck, nothing was ever said, unlike in a western cockpit. The Japanese would just stare out the window with sweat running down their brow and say nothing.

They used to shit me to tears with the crew meals when I was flying with an all-Japanese crew on the freighters. The airline would load on one western meal and the rest would be Japanese food. But one of the Japanese crew would always steal the western meal and leave a Japanese meal for me. I've always been quite fussy with food, and the Japanese airline food was far from gourmet sashimi, so I was never too impressed when this occurred.

When I started flying on passenger flights for Japan Airlines, things didn't improve much on the food front. On one occasion all three of us on the cockpit crew were westerners, so, knowing the crew meals would be suspect, we bought McDonald's in the terminal before departure and smuggled our contraband onto the aircraft. Once we reached cruise, it was time to scoff it, but just before we got it out, one of the Japanese girls came up to ask us if we wanted anything. The first officer quickly said that we would each like a full order of McDonald's (pronounced Mac-donarld in Japanese). The poor girl covered her mouth and giggled and said, 'Sorry, captain. We have no Mac-donarld on board.' We sent her packing. We called her back ten minutes later just as we were finishing the Mickey Ds. She let out a 'warrrrr' for the ages. It was gold.

And then there was the partying in layover ports. The Japanese pilots would always come out or invite you out. They knew how to have a good time, and some had no filter at all. But the next day when you said, 'Toshi-san, that was the best night out and you sure made a fool of yourself running naked through the lady's section at the *onsen* [Japanese bath house],' they would look at you with a blank look and pretend it never happened.

One of the things I noticed with the Japanese was that they were desperate to be liked and accepted by us. One of our American flight engineers, Bob, went on a trip to Frankfurt with an otherwise all Japanese crew. They had three days off in Frankfurt, so they hired a big BMW and went for a drive. I can't remember if it was intentional or by accident but at some point, fate put them outside the Nürburgring racetrack.

At the Nürburgring, you are allowed to pay to take your car on the track and do what is known as tourist laps. So they did, and why wouldn't you? On the third lap, as confidence started to exceed ability by a significant amount, they rolled the car on a corner, and it ended up on its roof, slowly spinning. Bob told how it still hadn't even stopped spinning and the Japanese were all out of the car posing for photos in front of it.

The story doesn't end there, because, as you can imagine, racing a car on a racetrack is 'verboten' in the small print of a German insurance policy. News of this all reached Japan Airlines HQ in Tokyo about 30 seconds after it happened (the car was still spinning then too). Bob was worried that he might be fired, and when he got back to Tokyo, he was invited to attend an interview with the big boss over green tea and seaweed biscuits.

As Bob stood to attention, the boss asked if he had any photos, which Bob duly presented. Then the boss said, 'Congratulations. Very nice to see foreign crew enjoying time with our Japanese crew.' And that was that. The crew members involved all had to fork out for the damage, but at least they kept their jobs.

The Japanese pilots were subject to many restrictions that western pilots are not. They were only allowed to fly to certain groups of airports so that they became experts at those airports. This was a company requirement to raise safety levels. In general, western pilots can go anywhere as long as there is a runway. The Japanese pilots were scared of many of their trainers as well. Trainers and checkers were treated like God. Loss of face was also a big deal for the Japanese. In my humble opinion, all these things only served to undermine their confidence a little, which is a shame.

At the end of August 1994, while I was working for JAL, I had been moved from the Anchorage base to the Honolulu base. It was announced that we would be starting to fly to the brand-new Kansai Airport near Osaka. The airport had been built on an artificial island in Osaka Bay 3.7 km offshore. It was due to open on 4 September.

Unfortunately, the underwater clay that the island had been built on was not ideal, and even by opening date, the island had sunk to the point where the runway had a downward slope, landing to the south, until about the 3,000 ft mark, then it levelled out. The island had been predicted to sink a maximum of 5.7 m over time, but by 1999 it had already dropped by 8.2 m, so by 2050, it may be better suited as a submarine base.

But getting back to the story, the first JAL international flight into Kansai on 4 September would be a flight from Honolulu. This must have created some stress for the Japanese pilots, because if you hadn't been somewhere, you couldn't go. Their answer was to make me the captain and have a JAL captain in the co-pilot's seat. They also rostered on another two JAL check captains to observe. None of them was willing to risk embarrassment by actually flying, so they all decided to watch me screw it up and suck wind through their teeth in the background.

So, off we went. Nine hours later, on the morning of the 4th, we arrived in the vicinity of Kansai. The landing would be towards the south. Because Kansai was planned for 24-hour operations, the authorities had designed an approach that flew a huge clockwise semicircle around Osaka Bay. They had also put many awkward descents and level sections in the approach, possibly as a joke. In the classic 747, this would be very difficult to fly accurately. These days the systems would do it all for you, but not back then.

So, there I was, working my arse off after being up all night, flying into a strange airport, being watched by three check captains. I can't remember it being much fun, but we survived, and everyone got a tick in the box. My landing probably drove the island an inch lower. According to Boeing, the correct landing procedure is to touch down fairly firmly and start the slowing down process without delay. Of course, ego precludes this, but in the case of my landing into Kansai, it was a good standard Boeing and JAL firmness technique, so what's not to like?

In another story about how severe some of the Japanese checkers were, one of our German flight engineers was doing his final check into Europe, and the Japanese check engineer was riding him hard. Finally, it all got too much for Gunter, and he turned to the checker and said, 'If you shove a broom up my arse, I'll sweep the flight deck as well.' This brain fart cost Gunter two months' more training.

My favourite JAL story actually happened after I returned to Qantas. We were flying JAL flights in Qantas aircraft from Nadi in Fiji to Tokyo and back. For this we had all Japanese cabin crew. The JAL standard procedure was for all crew to meet in the first-class section after boarding the aircraft for introductions and a crew brief.

The captain would run this brief and start by introducing the rest of the cockpit crew. Having worked with the Japanese for a few years, I understood that, although their command of the English language was very good, they could easily miss the subtleties of what was being said. With this in mind, I introduced first officer Ted of Habeas Corpus, second officer Graham of Uranus and engineer Ron of Great Flatulence. Everyone smiled and bowed and carried on like all was in order, but my crew were struggling not to laugh. The best was saved for last, however, when the chief purser leant into Ron and said, 'Ron-san. Where is Great Flatulence?'

FIVE

WHAT CAN GO WRONG

READ WITH YOUR EYES CLOSED.

FEAR OF FLYING

Just between you and me, pilots aren't the suave, fearless heroes we make out to be. I have some phobias that I'll admit to, such as arachnophobia (fear of spiders), ophidiophobia (fear of snakes), galeophobia (fear of sharks) and Australiaphobia (fear of all three: Australia is teeming with them).

Fortunately I don't have a fear of flying, but up to 40 per cent of passengers get some form of anxiety about air travel. Also known as aviophobia, around 2.5 per cent of the population have been clinically diagnosed with the condition by a mental health professional. Not only will aviophobia stop these people from flying, but they, in turn, will convince their families and friends not to as well. So, anything that can reduce the effects of this will mean more people flying, which is good for them and good for the industry.

People with aviophobia experience persistent and intense anxiety when they travel by air or think about flying. Symptoms may include sweating, shaking, nausea, chills, clouded thinking, increased heart rate, shortness of breath, irritability, gastro upset and flushed skin. Some people experience full-blown panic attacks.

Other phobias such as claustrophobia, a fear of heights, and social or germ phobia can contribute to aviophobia. For some people, the anxiety starts weeks before a flight, and for others, it begins just before boarding the aircraft. Aviophobia costs some people their jobs and even their relationships, so it's a big issue. It all sounds quite unpleasant to me.

We've all heard the statistics about how much safer you are to take a flight rather than driving your car, and these days, it's usually cheaper and definitely a whole lot quicker to fly. But as you can see, for some people, the decision isn't quite as simple as that. Why do you think some passengers sit in airport bars drinking at 9 am while they wait for their flights? It's not because they are thirsty. It's a big problem.

Fear-of-flying courses are available online or in live group sessions. Some of the courses are run by airlines and others by private companies. In the live sessions, customers possibly get to meet pilots, talk about airline safety and may even get to board a plane. Courses generally start by convincing fearful flyers that their fear is based on false assumptions. Once this is achieved then education is the key. The more people understand about flying, the less they will worry as there will no longer be any nasty surprises.

The courses also teach relaxation and coping techniques. Relaxation techniques can go a long way to reducing anxiety.

In the future, aviophobia might become less of an issue. In an ideal world, you would turn up for your flight, get put to sleep and be loaded into a capsule with a hydrating drip in your arm. Then you'd be woken at the other end and sent off afterwards with a couple of sandwiches and a coffee, feeling fully relaxed. You might even choose to be 'valeted' in flight. What's not to like about that concept? But, sadly, that's not a reality yet.

AVIATION MEDICINE

I'm not a doctor or a scientist, so the information in this section comes from my research or the fact that I've suffered from some of this stuff. If you think these topics might apply to you, then best you get professional advice. Any complaints about this content directed to me can be put in the special bin marked 'Waste' over by the door.

JET LAG

Jet lag happens when your natural circadian rhythm is altered because of travel to a new time zone. Your circadian rhythm is your internal clock that your body uses to manage sleep and wake times. The body clock uses daylight, your temperature and your hormones to achieve this. Travelling disrupts the body's ability to manage its internal clock, through the normal means.

The further you go west or east, the more it affects you. For example, if you fly to Europe, you will be nine or so hours behind Australia. This means you'll feel like going to bed really early but will wake up hungry at 2 am.

The best way to adjust the body clock is exposure to sunlight. In general, it takes a day to adjust by an hour, so when you go to Europe, it's at least a week before you are over the jet lag. In my job, because of flying east and west all the time, sometimes the body completely loses track of where it is, timewise. You don't really know whether you are Arthur or Martha. It can't be good for you long term. I got to the point where I would sleep when I was tired and eat when I was hungry. What more could I do?

The media will have you believe that jet lag is caused by time on a plane, and they like to talk about how this sports team or that will be suffering from jet lag. That's just BS. Time on a plane may dehydrate you, or leave you exhausted from finding it impossible to sleep in economy class chairs, but it does bugger all else. In the case of Tokyo, which is only one hour behind Sydney, there is hardly any jet lag to speak of, even though the flight is nine hours long.

Jet lag is thought to be worse as you travel from west to east, and you may also be more susceptible to jet lag if you travel frequently and also if you're an older person.

So, what's the secret of dealing with it? When you arrive at your destination, try to get straight into the new time zone and forget about your old one. If you eat and go to bed according to the time at your destination, you will get in sync with the new time zone quicker. When your plane is in the air, try to sleep during your destination's night-time. You should also avoid the urge to sleep when you arrive if it's daytime. This can make it difficult to sleep at the correct bedtime. If you force yourself to stay up on that first day, you will be a lot better off.

Exposure to sunshine can wake up your body and reduce the release of melatonin hormones that make you sleepy. Many aircrew have found that taking up to 5 mg of melatonin at bedtime in the destination helps to go to sleep and reset your body clock.

INFLIGHT RADIATION

Cosmic radiation is a much-discussed topic when it comes to crew health, so it is also a consideration for passengers who fly relatively frequently too. From around 26,000 ft, the ambient radiation increases by about 15 per cent for every 2,000-ft increase in altitude. And because the Earth's magnetic field is weaker at the poles, far more radiation exists in polar regions.

This is the case no matter the time of day or night. Multiple studies have been done over the years, and it is considered that the risk to aviation workers and passengers is insignificant. However, if your flying mainly involves long, high-altitude polar flying, you may want to investigate further.

Women who are pregnant need to be warier and should consult on this as early as possible. There are rules in place about female aircrew flying while pregnant, so it must be a consideration for passengers as well.

Airline pilots do suffer an increase in melanoma rates of two to six times the average population. This is thought to be related to UV radiation, however.

When I was a second officer, I was based in Frankfurt, Germany, for two months, shortly after the Chernobyl disaster. This was a tense time. One of the engineers carried a portable Geiger counter to keep track of his radiation dose so that he could sue the company later. Interestingly, there was more radiation at altitude away from Europe than there was on the ground in Frankfurt.

When I was working for Japan Airlines, we westerners discovered straight away that the flights were all being planned at relatively low altitudes. It turned out that the JAL pilots' union had imposed rules on maximum altitudes due to radiation exposure. So instead of sitting in the jet stream where the tailwind was the strongest, they would sit lower with less tailwind and more turbulence on the jet stream fringe. We foreigners would ignore the flight plan and, when light enough, we'd climb into the jet stream core and sit on minimum airspeed, letting the wind do the work. We could easily save 5–7 t of fuel on a flight by doing that, and my new third ear has made up for my general hearing loss over the years.

Later when I was back at Qantas, I flew into Tokyo just after the Fukushima disaster. The company told us that it was completely safe, but to be cautious, we should remain on the aircraft. Since then, I've grown a fourth ear, which has made me symmetrical again.

DVT

After the press started making a big deal of deep venous thrombosis, everyone suddenly discovered that they suffered from it. It became known as 'Economy Class Syndrome', which isn't really the truth. But why let the truth get in the way of a good story? As a result, nearly everyone in economy class wanted a free upgrade to business class. Let's take a look at what DVT is.

DVT is a condition in which one or more blood clots form in the deep veins of the legs. If blood is allowed to stagnate, clots may form in the veins.

If the blood clot dislodges, you have a real problem. The clot then travels in the bloodstream, and lodges in other veins or arteries, causing a blockage. This blockage is called an embolism. It can be life threatening, especially when the embolism occurs in the brain, heart or lungs. A pulmonary embolism (a blood clot in the lung) is the most common of these serious DVT complications.

Risk factors for DVT are many, and include pregnancy, contraceptive-pill use, blood clotting disorders, a previous or current history of cancer, a personal or family history of DVT and recent major surgery.

There is no specific evidence that air travel adds to the risk of DVT relative to other forms of travel. And business class, first class and aircrew are not immune from DVT.

Recommendations to reduce the risk of DVT include:

Walk around the cabin regularly.

Drink plenty of non-alcoholic, non-caffeinated fluids.

Recline the seat when practical.

Flex the calf muscles frequently.

Consider wearing medical compression stockings.

Delay travel (Maybe until your next life when you can afford first class).

A close captain friend of mine, who is known for his quick wit and intolerance of fools, told me once about an encounter with a DVT upgrade opportunist. He said that the customer service manager had come to the flight deck and mentioned that there was a particularly annoying woman in economy who was demanding an upgrade because of DVT. She was refusing to back down and was annoying everyone around her. She was demanding to see the captain.

So, he went down and could spot her a mile away. Here's how the conversation went:

'Good morning, madam. You asked to see me. Is there something I can help you with?'

'Yes, captain. I've asked for a move to business class because, as you can see, I'm middle-aged and slightly overweight so am at risk for DVT.'

'But, madam, you've just described half the people in the economy section. We can't upgrade them all.'

'But, captain, DVT runs in my family.' At this point my buddy said he wanted to say, 'No one runs in your family.' But he resisted and instead said, 'I'm sorry, madam, but I can't upgrade you.'

'Do you know who I am?' she asked. 'I'm very good friends with the chairman of the board.'

'Ah, Olivier Goudet. A fine gentleman,' said my buddy.

'Yes,' said the passenger, 'we went to school together.'

Now to fill you in here, my buddy had just been reading an article on the business side of Krispy Kreme donuts and Olivier Goudet was chairman and CEO. So he said, 'Well I can see he's been sending you plenty of product.'

'What do you mean?' she demanded.

'Olivier Goudet is chairman of Krispy Kreme donuts, and it looks very much like you've been eating way too many of them. There will be no upgrade. Now, stop bothering everyone. Goodbye.'

'Well, I never,' she said.

As my buddy walked away, he saw everyone around her smirking. He never heard any more about the incident.

HYPOXIA

Hypoxia refers to a lack of sufficient oxygen in the body, which can impair the brain and other organs. Hypoxia occurs due to reduced barometric pressures encountered at high altitudes. While the concentration of oxygen does not change at higher altitudes, the lack of atmospheric air results in less available oxygen. This means aircraft occupants are at risk of suffering from hypoxia.

Despite technological advances in aviation, hypoxia still occurs today. There have been a couple of notable incidences in recent years of corporate jets flying until they ran out of fuel, then crashing. These incidents have been put down to hypoxia. These accidents would probably have happened because the pilots were unaware that their cabin altitude had climbed to dangerous levels.

Symptoms of hypoxia include restlessness, headaches, confusion, rapid heart rate, rapid breathing, anxiety and blue skin (cyanosis). There are different stages of hypoxia depending on how high the cabin altitude gets, but under normal circumstances, there should be minimal effects in an aircraft cabin. Passengers with existing heart and lung disease seem to suffer the effects more. It is possible for healthy people to acclimatise to low levels of oxygen. A good example is those who spend time at base camp for Mount Everest acclimatising before attempting to summit. Some have even made the whole climb without supplementary oxygen.

BLOCKED EARS

Normally the air pressure inside the inner ear and the air pressure outside are essentially the same. Even if you were to hike to the top of a mountain, the slow speed of your climb would allow time for the pressure to equalise along the way. Blocked ears occur only when the change in altitude is too rapid, as in air travel, and the inner and outer pressures don't have time to equalise. This is known as ear barotrauma.

It can happen while climbing after take-off but is more likely on descent. It can be very painful and can result in ruptured eardrums. It's particularly bad when you are congested with a cold. Babies have it worse as their eustachian tubes are narrower than those of adults. That's why you will generally always hear a baby screaming somewhere on descent.

One time on descent into Singapore from Europe, one of our flight attendants had a head cold. She was having trouble with her ears. The flight service director brought her into the forward crew rest area. Suddenly, she let out a bloodcurdling scream as her eardrum burst. The scream was so loud as to be unforgettable for the other crew who heard it. She was seen to by the Qantas doctor in Singapore, where she was grounded and eventually put on a ship back to Sydney.

CABIN AIR QUALITY

You may be surprised to hear that the air inside a plane is cleaner than you might think. Thanks to HEPA (high efficiency particulate air) filters and efficient circulation on commercial aircraft, the air you breathe inflight is much cleaner than the air in office buildings, bars and your own living room.

Modern commercial aircraft are equipped with HEPA filters. That means that the airflow actually mirrors the constant streamlined airflow of an operating room with minimal or no crossover of air streams. Air is pumped from the ceiling into the cabin at a speed of about a metre per second and is then sucked out again below the window seats. About 40 per cent of a cabin's air gets filtered through the HEPA system. The remaining 60 per cent is fresh and comes in directly from outside the plane. While the aircraft is cruising, on average, the cabin air is completely changed every three minutes.

Of course, this doesn't mean you are bulletproof. If you are sitting next to a sick person who is coughing or sneezing, chances are you will get sick as well. That's why people who are sick shouldn't fly. Apart from being bad for you and risking ruptured ear drums, it's just not fair on anyone else. When I am a passenger on an aircraft, this is the thing I hate the most and another reason to carry the shotgun.

CABIN SURFACE GERMS

If you are someone who likes to think that as long as you can't see something, it can't hurt you, then I suggest you close your eyes while reading this section. Germs on aircraft have been sampled many times over the last few years, and the results are not encouraging. Part of the problem is inadequate cleaning. Some airlines require flight attendants to clean the aircraft in between flights, and there simply isn't enough time to do it properly.

In one set of tests, pathogens, yeast and mould were detected on the majority of aircraft surfaces. The most contaminated surface was the headrest, followed by seat pocket, washroom door handle, tray table and seatbelt. The most concerning finding was the presence of E. coli bacteria, which can cause diarrhoea, vomiting and stomach pain, on the seat pocket and the headrest. This indicates faecal contamination, and it's understandable how it could get in a seat pocket from nappies, but to get it on a headrest is quite an effort. My advice is to try and forget this detail before your next flight.

Other sources of nasties on planes include bathroom taps and floors. Blankets are often reused multiple times in between cleans and often have mould on them. I could go on but it's making me feel ill, so I might leave it there. I'd advise taking some high-strength disinfecting wipes with you next time you fly. You might even be able to buy some from an enterprising cabin crew member, for a price.

Having discussed all that, it's important to remember that germs are everywhere. All forms of public transport would have similar levels. Humans grow somewhat immune as they age, so you'll probably be just fine if you don't think about it too much.

MOTION SICKNESS

For some people, motion sickness can be pure hell. It might surprise you to hear that I suffer from it. Once I got caught for 12 hours on a yacht in rough seas off Hawaii and death by drowning seemed like the best option at the time. On my first flight training to be a navigator in the air force, I was as sick as a dog with cold sweats, dizziness and vomiting. I almost decided aviation wasn't for me. Luckily, I stuck with it and built up a tolerance of sorts.

That lasted until I graduated and was posted to P-3B Orions. The navigator on an Orion sat in the middle of a dark tube with no windows and no instruments to show which way was up. We would zoom around at low level in bad turbulence, making turn after turn while tracking submarines and fishing boats. I never knew whether I was upside down or the right way up. To add to this excitement, the guy next to me used to light cigarettes and leave them smoking in the ashtray between us. For almost two and a half years I was continually airsick. I got so used to it that

I would fill a bag, ask the ordinance man to bring another and just get on with the job. It became routine.

So what causes motion sickness? Your vestibular system uses a combination of your inner ear, vision and your body's ability to sense movement and position. Dizziness occurs when at least one of these three misfires, or the combination is not working as it should.

Motion can mess up how these systems communicate with each other. For example, if you are sitting in the middle of an aircraft and can't see out the window, your eyes will tell your brain that you're still, because your surroundings are, *but* your body and inner ear can sense the movement and tell your brain this. These contradictory messages are what cause motion sickness symptoms such as dizziness and nausea.

Motion sickness pills can help, but remember that they need to be taken well before the motion occurs. Reading doesn't help. You should try and keep your head upright.

Looking out the window gives you a better chance of reconciling all the prompts you are receiving. I've always found I have less of a problem if I don't think about it and I'm busy. And of course, if none of these work, there is always the option of death to end it.

For the record I've never suffered from it as a pilot while at the controls, although I did have to let the first officer fly once on approach into Sydney while I vomited from some evil bug I picked up in Jakarta.

PILOT MEDICALS

Airline pilots have to go through full medicals every 12 months. Once they reach a certain age, they are subject to stress ECGs as well if they are more at risk due to being overweight, having high cholesterol and smoking etc. At 60, pilots then have to have full medicals every six months. Authorities are very tense about pilot health and rightly so. Despite this, it's not unknown for a pilot to throw in the towel (drop dead) on a flight.

MEDICAL EMERGENCIES

Medical emergencies were something that arose from time to time right through my career. It was unusual to have two flights in a row where the customer service manager didn't have to call for a doctor onboard. Being a doctor on an aircraft would be painful. You could pretend to be a drain layer, but could you really ignore it if someone was in trouble? I couldn't. I have had occasions where doctors have been up all night helping, and when they walk off at the end of the flight, they've been given a bottle of cheap bubbles. They should get their ticket refunded.

Over the years, I've had to make a few diversions for medical emergencies. The company tries to stop really sick people from travelling, but that's not simple. There have been many cases of obviously ill people at the boarding gate insisting on continuing on a flight. 'But, madam, your head has exploded.'

'I'll be okay. It will grow back,' they reply in sign language.

Because airlines, rightfully, are wary of being sued by passengers, they have worked out how to offload the medical responsibility for looking after passengers. They use a service called MedLink. I bet it's not cheap. MedLink supplies inflight medical assistance and pre-flight passenger assistance by satellite phone. They are staffed by board-certified physicians and other professionals. If something goes wrong on a flight, there is no guarantee of a doctor being on board or that they will choose to identify themselves. So, to reduce liability, many airlines pay for the MedLink service.

Nowadays, even if there is a doctor onboard, Medlink is contacted and given the details. A Medlink physician then recommends treatment and whether a dash to the nearest airport is necessary. Ultimately the decision about where to fly the aircraft rests with the captain. Still, if you don't follow their recommendations and the passenger dies, you would need a very good reason why you didn't.

A medical diversion generally involves dumping fuel down to maximum landing weight, landing at a place with a hospital of repute, then refuelling and getting back on track. It is a costly exercise for the company because of the wasted fuel, the extra landing fee and the ongoing delays. Fuel dumping, or jettison, is required if an aircraft is above its maximum landing weight. For the first few hours of a flight, a diversion for a medical emergency will generally involve dumping. Although aircraft can land at a weight above the maximum landing weight, they will then be stuck on the ground until engineers have carried out lengthy overweight landing checks. So, if an aircraft needs to continue, fuel dumping must be used.

Larger aircraft have systems that can dump fuel out of the end of the wings, at very high rates, well in excess of 3 t per minute. As long as the dumping is carried out above 5,000–6,000 ft, jettisoned fuel will evaporate well before reaching the ground and shouldn't be an issue for people on the ground. If possible, fuel jettison is carried out over the sea or over non-built-up areas.

On 24 January 2020, we got airborne out of Dallas for Sydney. About two hours short of Fiji, while I was off on a break, someone decided to have a heart attack, or at least the beginnings of one. MedLink was called, and they decided that we needed to land as soon as possible. So, I was rudely awoken and called back to the flight deck. We were an hour out of Fiji by that stage, so we headed for Nadi.

We got organised for the arrival. We also worked out that, by the time we got through refuelling at Nadi, we could only make Brisbane before we would run over crew duty hours limitations. At that point, the second officer called Sydney and asked them to prepare a flight plan for Nadi to Brisbane.

After landing at Nadi, all the processes for departure began while the sick passenger was offloaded. Then the company called to ask why we weren't heading for Sydney. As previously advised to them, we couldn't make Sydney under the 20-hour maximum tour of duty. The company had made a big deal lately about how a diversion was required rather than exceeding 20 hours. So, we advised them that if they didn't hurry, we wouldn't make Brisbane either. At that point, the company conceded and started to produce the Brisbane plan, which then took another 20 minutes.

We got airborne out of Nadi as soon as possible after the flight plan arrived and started heading for Brisbane. But once we were on track, it became apparent that we wouldn't make Brisbane, so in accordance with company policy, we diverted to Auckland rather than back to Nadi. That was because better support would be available in Auckland. Of course, the company wasn't happy, and neither were the passengers, who were stuck in Auckland for eight hours. But that's life. I heard later that our man we offloaded in Nadi had died.

It's not uncommon for people to die on aircraft too. This is inconvenient, both for them and the crew. If there is no one onboard who can legally certify the death, then the victim can't be assumed dead. In that case, resuscitation attempts must be continued, and a diversion is mandatory to the nearest suitable airport.

If a there is a doctor onboard who can certify a death, then there is no need to divert. In theory the flight can then continue onto its destination. This raises issues about what to do with the body. On full flights, this can be a real problem. Sometimes the only option is to leave them where they are and put a blanket over their heads. This is quite disconcerting for those seated around them, but better than just putting dark glasses on them *Weekend at Bernie's* style.

On one occasion, on a flight from Sydney to Honolulu, a death occurred around two hours before landing. It was during a breakfast service and an elderly lady alerted crew that her husband was not responsive. Crew immediately moved him to a space near the doors and three of the crew took turns providing CPR. Unfortunately, no doctors were onboard. This meant that the three crew had to continue CPR until the passenger could be pronounced dead by either a doctor or a paramedic.

Some passengers complained that the normal second coffee service had not been served. This was due to the three crew members being unavailable due to CPR requirements. The three crew members continued CPR during landing.

Even though US officials had been notified of the situation by the pilots prior to arrival into Honolulu, after landing, there was a delay of around three hours due to the paperwork required, because the deceased man was a British passport holder who died on 'Australian soil' (an Australian-registered aircraft is always considered Australian soil).

Finally, passengers were allowed to disembark, and the outbound crew were allowed to board to relieve the inbound crew. The wife of the deceased passenger remained onboard. An ambulance arrived, and paramedics came aboard from the rear left hand door. The aircraft was a 747-200 series, which had a 'shaving station' immediately behind the last row of the middle seats with a bulkhead just near the two sets of toilets. This 'shaving station' became a hindrance in the removal of the deceased. Once the man was placed on the stretcher and brought to the rear of the aircraft, they had to upend the stretcher, with the man's head down, to get around the corner of the middle bulkhead and clear the shaving station. All this was observed by the now very distressed wife.

On another occasion, on arrival into London on QF1, an elderly British lady, who had finally completed the long flight, disembarked the aircraft and was walking along the aerobridge when she suddenly collapsed. Staff were unable to revive her, so paramedics were called and pronounced her dead. It was later found that the cause of her death was due to deep vein thrombosis.

Of course, deaths are rare compared with other medical emergencies. On one flight, a middle-aged male passenger suffered an epileptic fit. It was on a flight to Sydney, during collection of the meal trays after the breakfast service. A passenger alerted crew, saying there was a man who had collapsed near the middle toilets on a 747-300. Crew got to him quickly and he was still seizing. They placed pillows and blankets around him so he wouldn't hurt himself against the toilet doors. More crew came but there wasn't much they could do except get gawking passengers away. The flight service director had been made aware and promptly made a PA asking if there was a doctor onboard. Luckily there was, but by the time he arrived on the scene, the passenger had recovered and had been seated in a row of spare seats. He was totally bewildered. Unfortunately, he didn't speak English.

The doctor took over and gently gestured to him, asking if he could take a look at him. Our flight service director came down with the inflight physician's kit, which is only used by a qualified doctor. The patient's blood pressure was taken and appeared to be low. The doctor believed he would be okay, but he checked on him often for the remainder of the flight. One crew member stayed with the passenger and tried to keep him calm. He obviously had no idea what or why this was all happening. The crew member eventually got him to show them his passport and found out he was from Bulgaria.

The pilots, who had already been informed of the medical situation, radioed the relevant authorities. After arrival into Sydney, a medical team came onboard and a ground handler, who was Bulgarian, came to interpret. It turned out that this passenger had never had a seizure before. It was decided by the doctor and the medical team that he be taken to hospital in Sydney for further assessment.

On top of medical emergencies are cases of unusual behaviour for crew to deal with as well. In one case, an older lady was involved. Crew heard a number of passengers yelling near one of the exit doors. They rushed to the commotion and found a woman trying to open the exit door. One crew member calmly went over to her and asked what she was doing. Another crew member explained to the very concerned nearby passengers that it was impossible to open an aircraft door inflight. Meanwhile, the offending lady told crew that she had had enough of this supermarket and wanted to leave! The crew settled her back in her seat, convinced her that oxygen would help to make her feel better, and so it did.

On another flight, an elderly gentleman in economy class, on his way to Sydney, was seated in one of the middle seats of a 747-300. He asked a crew member, in a very heavy Yorkshire accent, if they could help find his 'bus'. The crew member made an assumption that he was looking for a toy bus that might have fallen out of his in-cabin bag. So, the crew member went down on all fours searching under the seats around him for a toy bus. Another crew member walked past and asked the original one what they were doing. After an explanation, he decided to go over to the opposite aisle and search there. So, there they were, rummaging around on the floor, when another crew member came by and made the same enquiry. After receiving an explanation, he asked the passenger what kind of bus it was. His reply was, 'It's a big one. I have passengers. I need to find my bus.' The crew immediately called for oxygen.

THE WEATHER

The weather is something that pilots have to keep an eye on all the time, even if you are flying above it for a lot of the time in the cruise. It comes in many different forms, and a very good knowledge of the causes of weather is mandatory for pilots. Modern aircraft are extremely well designed to cope with any sort of weather, but avoidance, when possible, is always best from a passenger comfort perspective. Let's look at the different types of weather which affect aviation.

WIND

Wind has a significant effect on aircraft operations. Aircraft create lift from the number of air molecules passing over the lifting surfaces. This is also known as dynamic air pressure. As an aircraft climbs, the air density decreases due to an increase in altitude and a decrease in air temperature, but the aircraft still needs the same dynamic pressure to fly. So, the true airspeed needs to increase as the aircraft climbs, just to maintain the same number of molecules passing the wing (the desired dynamic pressure). An example would be that with a dynamic airspeed of 530 kph, the true speed through the air would be need to be around 925 kph at 35,000 ft. At cruise, this true speed through the air would be equal to the speed across the ground if there was no wind. Are your eyes starting to glaze over yet?

But there is rarely no wind. Sometimes the wind speeds are very high such as in jet streams. Jet streams are caused by the meeting of two large air masses of different temperatures or densities, with the wind arising from the difference in pressure between these masses. Due to the Coriolis effect of the Earth turning on its axis, this wind flows along the line where the two air masses meet.

Jet streams are high altitude air currents which usually flow from west to east at speeds of up to 220 kt or 400 kph and can move around a lot with the air masses which cause them. Airlines like to take advantage of the tailwinds going eastwards and avoid the headwinds in the other direction. Computer modelling helps with route planning and it's quite rare to fly exactly the same route two nights in a row.

Jet streams have a significant effect on the operation between Japan and Hawaii during the northern winter. For example, on one trip to Osaka from Honolulu, we took 9 hours and 55 minutes, and on the return the following evening, we took 5 hours and 55 minutes. How this works is that if you are flying directly into the wind, your speed across the ground will be your true airspeed minus the wind speed, so in the above case it's 925 kph – 400 kph = 525 kph. Going with the wind, your ground speed would be 925 kph + 400 kph = 1325 kph. As the wind angle changes away from direct on the nose or on the tail to more side on, the effect reduces, which is basic trigonometry.

Near the ground, wind that is not aligned with the runway direction is known as crosswind. This makes the normal process of landing more difficult. In the A380, the crosswind limit on a dry runway is 40 kt, which equates to 74 kph of wind trying to blow you sideways off the runway while you try to land. When it's raining and the runway is wet, these limitations reduce quite a bit to around 25 kt, depending on the airline. That can be quite restrictive, and sometimes flights are diverted or cancelled because of it.

Any component of headwind is welcome, because it means the ground speed of the aircraft is lower than the air speed, and it will take less runway to take-off and land. Conversely, tailwind is not something pilots like to be involved with unless it is light and there is no alternative.

Strong winds at low level can mean turbulence. If the wind passes over hilly terrain or substantial buildings near the airport, it can affect the aircraft's flight path considerably. One time on the 747-400, we were heading to Hong Kong while a typhoon raged there. Most of the flights heading there were diverting into alternates on the Chinese mainland or into Manila in the Philippines. I decided it was worth having a look as some flights were still getting in, so we made our approach in crappy conditions. The old Hong Kong Airport would have been a real drama in these conditions, and the new one was supposed to make life easier. But the new airport was built behind the steep hills of Lantau Island, so when you got wind from the east or southeast, it would cause severe turbulence and windshear, or large wind fluctuations, on the approaches to the airport.

On this occasion, it was just like that. There was heavy rain and severe turbulence, and windshear. On approach, the stall warning went off a couple of times. I managed a pretty good landing under the circumstances and taxied in. After shutdown, the crew all said, 'Nice work, boss.' I said, 'Really? I'm wondering why we were there at all. It might not have been my best management decision ever. So, nice work? Maybe not.'

Windshear is a difference in wind speed and/or direction over a relatively short distance. Atmospheric windshear is normally described as either vertical or horizontal windshear. Vertical windshear is a change in wind speed or direction with a change in altitude. Horizontal windshear is a change in wind speed with a change in lateral position for a given altitude.

In the case of vertical windshear, an example would be that at 500 ft on final approach, you had 30 kt of headwind, but at 200 ft only 10 kt of headwind. That's a loss of airspeed of 20 kt in 300 ft (around 20 seconds), and it can cause the aircraft to land short of the airfield, if not handled well. In the case of horizontal windshear, wind could be coming out the bottom of a nearby thunderstorm. As you approach the storm, the wind could be blowing 40 kt of headwind, but as you pass the storm, the wind is now blowing at 40 kt from behind. That's an 80 kt drop over a short distance.

Windshear has been known to cause many aircraft accidents, so pilots are well trained to avoid it in the first instance and deal with it in the second. Aircraft systems and procedures are set up to handle moderate windshear. Modern aircraft have predictive windshear systems that warn pilots in advance if windshear conditions exist. If windshear is actually experienced near the ground, it is treated like a life or death situation. There is no second chance if you underestimate it.

TURBULENCE

Given that the 40 per cent of passengers who are uncomfortable with or scared of flying are most concerned about turbulence, pilots should avoid it when possible. In the cruise, turbulence can be caused by four things. They are thunderstorms, jet streams, mountains and wake turbulence. Mountain ranges that are high enough and wide enough can deflect wind upwards and cause mountain waves and turbulence. An example of this would be the Rocky Mountains in the USA or those around the new Hong Kong Airport.

One of the problems with using the jet stream to push the aircraft along is clear air turbulence, caused by the friction that comes from higher and higher wind speeds flowing next to each other. It can get quite bad, and the worst thing is it can't be seen in advance. But it is predictable, and sometimes it's the price you have to pay for a quick flight. Going into the wind is a different story. It doesn't make any sense to get beaten up plodding directly into the jet stream. Sometimes it's smoother and more efficient to stay down low beneath the jet stream or fly the long way around to avoid it.

The worst clear air turbulence I ever experienced was on a flight from Tokyo to Taipei. In the winter, the jet stream always tends to sit over this route. Luckily, we were ferrying an empty 747, so there were no passengers or cabin crew. The turbulence got so bad that we couldn't read the instruments in front of us, because the cockpit was shaking so violently. At one point we got such a jolt that I suddenly remembered what I was in my previous life.

Wake turbulence is a disturbance in the air that forms behind an aircraft in flight. It includes a number of elements, the most significant of which are wingtip vortices and jet wash. Wingtip vortices are quite stable and can remain in the air for up to three minutes after an aircraft passes. The strength of wingtip vortices is determined by the weight and airspeed of the aircraft and, to a certain extent, the design of the wing. Wingtip vortices are the most dangerous component of wake turbulence.

Smaller aircraft passing behind larger aircraft need to be aware of wake turbulence risks. As the wingtip vortices are spinning tubes of air, much like a tornado on its side, they can flip a smaller aircraft over in an instant. Even a large aircraft passing through the wake turbulence of another large aircraft will be shaken up by it. One time when I was flying a 747 into Melbourne, I was following another 747, maybe 15 km ahead. The autopilot was controlling our aircraft, and suddenly the left wing dropped to a 35-degree angle of bank. I had to disconnect the autopilot to manually level the wings. That was caused by wake turbulence.

THUNDERSTORMS AND TURBULENCE

Different airlines have different policies on turbulence avoidance. At Qantas we would avoid moderate turbulence and thunderstorms, even if it meant diverting 100 km off track. This adds costs for the airline, but it's non-negotiable. As a result, my aircraft was only ever struck by lightning twice in 37 years. A friend of mine who flies for another airline said he had been struck over ten times in his career. Maybe he needs to go to confession more often.

Modern aircraft are equipped with good-quality weather radars. They show areas of precipitation in different colours, from green at the lower end to magenta at the extreme end. The weather is displayed on the navigation displays in the cockpit, so avoiding it is just a matter of pointing the aircraft away from the red and magenta areas on the screen. The areas of higher precipitation are indicative of water, ice, and updrafts and downdrafts. These are the areas where the worst turbulence will be.

These extreme areas are usually associated with thunderstorms. The other problem with thunderstorms is that they cause lightning.

LIGHTNING

Lightning can strike aircraft many kilometres from the actual storm cell, so pilots try to avoid storm cells by 15 to 20 km laterally if they can. Sometimes that's not possible and the pilots need to pick the best path of all the bad choices. That's when the seatbelt sign will be turned on in advance so that the cabin is well prepared.

Being struck by lightning in an aircraft, although unpleasant, isn't the end of the world. Aircraft are designed for it. A lightning strike will move through the outer shell of the aircraft, safely away from electrical wiring. The only thing you will hear is a noise like a hammer hitting a tin can, and the aircraft may shake a little as well.

Everything from flight computers to fuel lines to even the entertainment screens are isolated and protected from lightning. The composite structures that aircraft manufacturers use these days to save weight give the fuselage lower overall conductivity, however, so manufacturers lay unnecessary conductive wires in the body of the plane. While they are not wired in, they are 'attractive' enough for electrical current to pass through ('Ooh. Look at the arse on that cute conductive wire').

Additionally, airline fuel has been better engineered to be less flammable and not spark under stormy conditions. Fuel tanks are heavily insulated anyway. On top of this, modern aircraft like the B787 and the A350 have fuel tank inerting systems, which feature air-separation modules to generate nitrogen-enriched air that is used to reduce the flammability of fuel vapour in the tanks.

The new F-35 fighter jet does this too, but it doesn't work. The issue is with the F-35's OBIGGS (onboard inert gas generations system), which pumps nitrogen-enriched air into its fuel tanks to inert them. This prevents the aircraft from exploding if it is struck by lightning. Apparently, the tubing and fittings for the F-35 system cease to function over time due to various factors. At the time of writing, F-35s, which have been ordered by multiple countries at US$80 million a pop, are not allowed to fly within 25 miles of a thunderstorm. Ironically, the name of the F-35 is the 'Lightning'.

In Australia, which is a favourite holiday destination for thunderstorms, the military have ordered 72 F-35s. An advantage of all this silliness is that it makes it much easier to predict an attack. The enemy will only come in thunderstorm season and only at times of high humidity. Knowing this doesn't tell you how to defend against such an attack, but at least it wouldn't be a surprise. Perhaps the rules of war can be amended to say that attacks are not allowed in the wet season.

And now we return to our regular programming. Lightning strikes will leave small burn marks, and possibly even a small hole, on the aircraft skin at the point of the strike. The most inconvenient thing after a lightning strike is that the aircraft must be inspected by engineers before flying again.

On 18 April 2019, we were on the ground in Dallas waiting to take off with a full load of fuel. Air traffic control advised us of a line of severe thunderstorms stretching from 200 nm northwest to 500 nm southwest, and aircraft were having trouble penetrating them due to the height of the build-ups. We could see flashes of lightning in most directions to the west. Air traffic control suggested we wait it out on the ground. The problem was that if we did, then we couldn't go anyway, as we'd never make Sydney within our flight crew duty limit of 20 hours. So, I decided to get airborne and have a look. We would always have the option of diverting into Los Angeles or Honolulu on the way if the initial diversion used too much fuel.

We got airborne and started to track down the side of the line of storms heading between south and southwest right over the top of San Antonio. We looked hard for a gap in the weather to fly through, but none showed. Usually, two of us would have disappeared into the bunks for a rest, but I could see we had things to do, so we all stayed there. We basically made the bottom of Texas and crossed the Mexican border, heading at right angles to where we should be going, before we could turn.

More than an hour after getting airborne, we turned onto our new track following the Mexican border westward. We set the 'maximum range cruise' speed and sent information back to Qantas. We gave them a position, weight and fuel onboard, and asked for a new flight plan to Sydney. Eventually, we had a new plan to follow the Mexican border northwest until San Diego, then track over Hawaii, to the north of Fiji, over Noumea, close to Brisbane and arrive in Sydney with minimum fuel.

At that point, with everything sorted, I could leave the boys and go for a break. Against all the odds, we made it to the gate in Sydney. Normally this flight would take around 16.5 hours, but in this case the flight time was 18 hours and 3 minutes, and we covered around 16,000 km. That was a personal record for me.

No fuss was made. No thanks were received. It's what they pay you the small bucks for, and that's what's called 'getting the job done'.

ICING AND SNOW

Icing in flight isn't usually much of a drama at higher levels. It only occurs in visible moisture, or cloud, when it's colder than 10°C and warmer than −40°C. Most of the time in cruise, the temperature is colder than −40°C so any moisture is so dry that it can't stick to aircraft surfaces. In between the temperature ranges while on climb or descent, anti-icing is used. This involves pumping hot air bled from the engines into the areas where icing forms. On more modern aircraft, this is automatic. On older aircraft the pilots need to switch it on manually. Anti-icing applies only to wing leading edges and engine nacelles.

On the ground and near the ground, icing is much more of a problem. Not only can icing change the aerodynamic profile of the wing, but it also adds a considerable amount of weight to the aircraft and can lock up flight controls. A good understanding of, and strict adherence to, icing procedures is mandatory for pilots.

I had some limited exposure to these procedures when flying into Europe in the winter with Qantas, and we practised them often in the simulator. But when I was sent to Anchorage, Alaska, working for Japan Airlines, I was in for some shock treatment and fast learning when it came to cold weather operations in the 747. De-icing and anti-icing were a given. I remember ruining my good boots trudging around doing a ground inspection in a foot of snow mixed with orange anti-icing fluid on the ramp at Anchorage.

One of the rules for de-icing is for all surfaces to be clean for take-off. Initially the aircraft is de-iced with hot liquid. Then anti-ice fluid is sprayed on to resist further build-ups of ice.

Any sort of precipitation that occurs after the anti-icing has to be taken into consideration. Based on the conditions and type of precipitation, a 'holdover time' is calculated. If the aircraft is not airborne before that time, it needs to return to the bay and get the procedure done again.

Sometimes, we'd get held up in the queue or at the take-off point to the extent that we'd need to return to the ramp for another de-ice and anti-ice. It depended on the type of precipitation. If it were light snow, it'd be fine usually. Heavy snow, not so much. Freezing rain was a no-go at all generally. Airports with long delays in good weather would pretty much shut down in icing weather, because it was never possible to make take-off before your holdover time ran out.

Some airports that experience consistently cold winters are better prepared than others. They have de-icing bays that allow aircraft to taxi in and get de-iced just prior to take-off. Holdover times are much less restrictive on the course of operations at these airports.

After some weeks in the Anchorage winter, the airport taxiways would end up with a layer of compressed ice on them, so taxiing became a skill. It needed to be slow and, although we were never taught this, using asymmetric thrust and differential braking to coax the aircraft around corners was very useful. (Asymmetric thrust means using more thrust on one side of the aircraft than the other to help the aircraft turn. Differential braking would mean using braking on the opposite side from the thrust to assist the turn even further.)

We weren't allowed to park the brakes on the runway, which you might do on a clean runway if there was a take-off delay. The reason for this was that a Korean Airlines 747 had attempted to take off with the brakes on a couple of years earlier. They put on take-off thrust, which was enough to get the bird skidding off down the icy runway. About halfway down, all the tyres burst, and the aircraft just managed to stop before the end of the runway.

As you can imagine, Anchorage Airport was well set up for ice and snow, but sometimes there was so much of it that even they couldn't keep up, so they would work on keeping one runway open at a time. Sometimes even that was a big job. I took off there one time with the left-hand runway lights under snow and only the right ones visible. By the end of the winter, there would be huge piles of snow along the sides of the runways and taxiways.

Landing in these conditions was interesting as well. The ATIS (aerodrome weather report) would give a braking action report, starting at normal and reducing to poor. This was generally obtained from pilot reports. Braking action of poor or worse usually meant a diversion. On one landing I did there, the braking at the touchdown point was okay, but in the second half of the runway, it became apparent that zero braking was happening, and I had to use reverse thrust all the way to taxi speed.

Snow in itself is not a major drama. It's obviously not okay to have any on the aircraft surfaces for take-off unless it's a light dusting that will blow off during the take-off. It's when it melts and then freezes as ice that it becomes an issue. Fresh dry snow on the runway of a few inches depth doesn't cause much drag on take-off or landing. When it's hiding a layer of ice underneath, that is a different story. When I was living in Anchorage, I'd often go mountain bike riding on a layer of fresh snow. It was perfect. It was very grippy and created no extra drag to slow me down.

FOG

Fog in the winter is a double drama due to reduced visibility and icing. Freezing fog is a type of fog that forms in temperatures below 0°C. The tiny water droplets in the air remain as liquid. They become supercooled water droplets, remaining liquid down to temperatures as low as −10°C. This occurs because liquid needs a surface to freeze upon, like an aircraft wing. Temperatures below −10°C are cause for celebration in freezing fog.

The most common form of fog is radiation fog. It typically occurs on clear nights as the Earth's surface cools moist air immediately above it. When light wind is present, the chilled air is gently stirred through a deeper layer, thus forming a deeper radiation fog.

Valley fog is really just a type of radiation fog. It results from cold, dense air sliding down mountain slopes at night. The air collects in the valley floors, then forms as radiation fog. If there is an airport in the valley, the fog may not clear for hours after the sun rises.

Fog can also form when warm air moves over a cold surface. Warm air moving over cold ground in winter and sea fog drawn inland over a cool land surface are two examples of what is known as advection fog. A great example of advection fog is the fog that moves in off the sea at Los Angeles, often affecting LAX Airport.

Modern aircraft are capable of landing in fog in conditions as low as a cloud base of zero and visibility less than 200 m. Procedures to achieve this are known as Cat II and Cat III. Pilots need to be specifically trained to carry them out, and airports must satisfy minimum lighting and levels of backup equipment to offer these sorts of approaches. It takes a certain amount of trust from the pilots to allow the aircraft to fly to these low limits, because, in the case of Cat III, the only things a pilot sees at touchdown are a few centreline lights. Here's a story from the Anchorage basing that covers low visibility flying.

In early November 1993, the annual Beaujolais Nouveau absurdity started again. Beaujolais Nouveau, as the name would suggest, is a red wine produced in the Beaujolais region of France from Gamay grapes. It is released for sale annually on the third Thursday of November, after being fermented for just a few weeks. It's also famous for distributors racing to get the first bottles to different markets around the globe.

Because the Japanese bought into this weird tradition, it was considered very cool to spend a shitload of yen to drink rubbish red on a specific day of the year. So, every year leading up to that day, we'd fly freighter loads of Beaujolais from Paris to Japan via Anchorage. If you do the math, at 110 t of freight and 1 kg a bottle, that's 110,000 bottles per flight. Unfortunately, that's also 110,000 nasty hangovers.

And so it was that in the middle of November, I was scheduled to fly to Paris, with Big George Palfi as my first officer, to pick up a load of this plonk. But it also turned out that each new captain had to do a 'Cat II low visibility' check out before being cleared to fly to Cat II without supervision. So, we were joined by a Japanese check captain, who would do this check on me. In those days, we were automatically cleared to fly down to a cloud base of 200 ft and visibility of 600 m. But once we were checked to Cat II, we could then fly down to 100 ft and 400 metre visibility using the aircraft's auto-land system. So, the qualification would mean less likelihood of a diversion in crap winter weather going forward.

We launched for Paris with a pretty average forecast, but as we went over the North Pole, the weather in Europe got worse.

Paris was right on the Cat II minima of 100 ft cloud-base and 400 metre visibility, as were most main centres like Frankfurt. The only place that was any good was Geneva, which had a cloud base of 400 ft. This made it legally okay as an alternate destination for us.

With about two hours to go to Paris, the Japanese checker emerged from the bunk to check on progress. I briefed him about the weather, and he thought about it and said, 'But I'm not qualified to fly to Geneva.' You'll remember JAL captains needed to be checked into each destination, and most were limited to only a few destinations so that they got really comfortable with them. Then he said, 'I am pilot in command, so if I say we divert, we divert to Frankfurt.'

It would have been illegal and stupid for us to do that, given the weather in Frankfurt. The legalities of the navigation laws and plain commonsense far outranked the questionable local rules that some airlines have. When the checker went back to the bunk, I said to George, 'If we have to divert from Paris, we are going to Geneva, even if you have to take him down the back and tie him up.' George agreed wholeheartedly. I think he was kind of excited to be involved in a mutiny.

Just before the descent into Paris, nothing had changed and the Japanese checker jumped into the first officer seat for the arrival with Big George on guard at the back. I gave my arrival briefing, and in it, I briefed for Cat II automatic landing. Cat II is not allowed without auto-land, as the auto-land system is not subject to the human errors that are common in low visibility conditions. Humans are better off monitoring the autopilot doing its job.

So, I briefed for Cat II auto-land, and the JAL checker said, 'But you aren't qualified to do that yet. So, you will do Cat II manual landing.' At this point, I looked back at George, who had a wry smile on his face, and then I said, 'I'm not prepared to do that. So, it's either auto-land, or you can fly the approach.' The checker sucked through his teeth for a bit and then said, 'Okay, you do auto-land.'

I did, and the weather was as crappy as they had forecast. At 100 ft, I could just see the runway lead-in lights. In the end, the autopilot did a great job. We touched down, and I pulled the engines into full reverse. I had hardly got them into reverse, and we were still doing a good 110 kt, when the JAL checker started slapping me on the back and congratulating me. Needless to say, I passed and was signed off for Cat II.

RAIN AND SHINE

Rain is an issue if it's heavy. Aircraft are equipped with wipers but when the rain is heavy, visibility in general reduces, and it is much harder to see out through the windscreen. The view comes in waves as the wipers move backwards and forwards. Rain also makes the runway wet, which increases the landing distance required for an aircraft.

Hot weather tends to affect aircraft the most on take-off. If an aircraft is heavy, it will use much more runway in hot temperatures to get airborne than in cold. This is because the air in warm temperatures is less dense, and engine performance is also reduced in higher temperatures. Sometimes this actually restricts when aircraft fly. Take-off out of Johannesburg used to always be scheduled for after dark. Not only was the airfield at 5,550 ft above sea level, but the temperatures there were always higher than standard. We would have to wait until the temperature dropped enough to allow our take-off.

VOLCANIC ASH

There are two famous 'Rings of Fire'. The most well-known one is the morning after-effects of a good Indian meal. The second, which I'm going to talk about now, is a bunch of volcanoes that circle the Pacific Ocean.

The Ring of Fire is a path around the Pacific Ocean distinguished by active volcanoes and frequent earthquakes. Its length is approximately 40,000 km. It stretches from New Zealand in the southwest all the way around the Pacific Ocean in a clockwise direction to Chile in the southeast. It runs along the boundaries between several tectonic plates, and it contains 75 per cent of Earth's volcanoes.

Volcanic ash can be hazardous for aircraft. It can be abrasive for surfaces, including glass and windscreens. It can infiltrate fuel tanks and contaminate the fuel, block sensor tubes and damage engines, causing them to fail. It's worth avoiding.

Perhaps the most famous incident of this was British Airways Flight 9 in 1982. The aircraft flew into a cloud of volcanic ash thrown up by the eruption of Mount Galunggung around 180 km southeast of Jakarta, Indonesia. This resulted in the failure of all four engines. Because it was dark, the crew had no real idea what had caused it. They were able to get three engines restarted during the glide. They had to land the aircraft with minimal visibility through the windscreen, thanks to the sandblasting the windscreen had received. It was a supreme and commendable effort to save the aircraft.

In the subsequent investigation it was found that the dust in the ash cloud deprived the engines of sufficient oxygen from the atmosphere to maintain combustion. It also sandblasted the windscreen and landing-light covers, and clogged the engines. As the ash entered the engines, it melted in the combustion chambers and adhered to the inside of the power plant.

As a result of the investigation, new procedures for dealing with volcanic ash were created and adopted. The most pertinent is to reduce thrust to the minimum right away to cool the core operating temperatures of jet engines below the melting point of silicate ash particles (around 1,100°C). Idling the engine is the best way to quickly cool them. If this doesn't happen at or before the first signs of engine malfunction, and clear air cannot be found, engine failure is likely to occur. Of course, this means immediate descent, which can be a bit hectic in areas of high aircraft traffic density or poor availability of air traffic control communications.

Since that event, and even today, crews have been trained in the simulator to handle volcanic ash encounters. It is never easy, even having read all about it and knowing what is coming. This reinforces how well the British Airways crew did.

Nowadays, as a result of the BA 9 and other incidents with aircraft and volcanic dust, everyone is pretty tense about it, and airspace is often closed down as a result. In 2010 there were multiple eruptions of the Eyjafjallajökul* volcano in Iceland. Authorities were concerned that volcanic ash ejected during the eruptions would damage aircraft engines, so the airspace of many European countries was closed to commercial aviation traffic. At the time this was the largest air-traffic shut-down since World War II. The airspace closures caused millions of passengers to be stranded in Europe and across the world as all flights in and out of Europe were cancelled.

Anchorage is right in the middle of the Ring of Fire and is famous for its massive earthquakes and the odd volcanic eruption. While I was based there, I experienced two significant eruptions. The first occurred while I was riding my mountain bike in Anchorage. It left a half-inch layer of volcanic ash over everything. Volcanic ash is gritty and super fine. It can get into places even a fart can't penetrate. I was picking volcanic ash out of some very weird places for days afterwards. That eruption of Mount Spurr, in August 1992, shut down Anchorage Airport for over a week.

The second eruption happened while I was flying from Anchorage to Japan. Captain Jan Elberse, a Dutchman and senior pilot with Japan Airlines, had come to town to see how all us transplants were doing. He was a good friend of mine and had been to our place for dinner.

The next day, I was due for another line check to Tokyo with a Japanese checker, so Jan decided to fly back on my flight. The weather wasn't too bad when we took off, and it seemed like the flight would be pretty straightforward. Shortly after take-off, both Jan and the checker disappeared into the bunks.

---

* There will be a spelling test later.

About three hours into the flight, we got a call from air traffic control saying that a volcano had erupted on the Kamchatka Peninsula and that we would need to divert 500 km south to avoid the ash. The first officer and I determined that we could just afford to do this and still make Tokyo, so we diverted to the south. At the end of the diversion, we turned towards the west and found that the headwind was half as strong again as forecast. Then, 30 minutes later, the weather in Tokyo deteriorated to the point where we needed an alternate airport.

Suddenly we were looking like arriving in Tokyo with minimum fuel, in bad weather and with no fuel for an alternate. Because of the direction we were coming from, Tokyo was now the closest suitable airport, so there was no inflight diversion capability.

About an hour and a half out of Tokyo, the checker and Jan both woke, had a snack and a coffee, then joined us up front. 'Everything good?' they asked. I filled them in on the situation, expecting the checker to panic and suggest a ditching or something, but he just laughed and said, 'Haha. You've really got your work cut out for you now.' Jan was no help either. He didn't stop laughing for ten minutes. In the end, we made it with minimum fuel, and I did a landing that no one even felt, just to shut those two cheeky bastards up.

AIRCRAFT SYSTEMS AND INFLIGHT EMERGENCIES

A modern jet aircraft is a wonderful piece of equipment, which is testament to man's genius. When you compare an aircraft like a Boeing 787 or an Airbus 350 to the Wright brothers' aircraft and consider that there is only 120 years between the two, it's even more impressive.

In this chapter on the systems in modern aircraft, I don't want to get too bogged down in detail, because I don't want to have to keep waking you up. So for each system, I'll cover a little bit of history when relevant, some technical details, information on what can go wrong and how pilots deal with it when it does. My intention is not to scare you but to impress on you how safe you actually are.

On that note, it's important to remember just how safe aviation is. The reason for this is that modern aircraft are designed within the concept that no single point of failure should be able to cause an incident. Each critical system has multiple backups. Most of these systems operate in the background, and passengers will be generally unaware of their existence.

ELECTRICAL

Electrical systems on aircraft power just about everything from the inflight entertainment to lights to computers and electronics. On the Wright Flyer, the closest thing to an electrical system was the 'make and break' set-up on the combustion chambers in the engine. The engine didn't even have spark plugs. On a Boeing 787 the use of electrical power has increased to the extent that most systems which had previously been air driven have now been converted to run off electricity as well.

To power this, there are two generators on each engine, two generators on the auxiliary power unit, a ram air turbine in case the generators all fail, and an aircraft battery. That's pretty impressive. It would be extremely rare to lose all electrical power on a modern aircraft, and even if you did, the battery would still power certain critical systems. In my career, I have never seen anything more than a single generator failure.

HYDRAULICS

The Wrights didn't need hydraulics, as their flight controls were light enough to be controlled by the pilot. These days, with the size of modern flight controls, hydraulic assistance is required. Hydraulics are also used for flap and landing gear extension, with electrics used as a backup. Brakes are also hydraulic.

Hydraulic systems can fail, so there is always more than one system. There were four hydraulic systems on the 747. I lost a few hydraulic systems in my career. When I say lost, I'm exaggerating. I always knew where the systems were but was never quite sure where the hydraulic fluid had gone.

With the loss of a hydraulic system, it was normally just a matter of going through a procedure and Bob was your uncle.

But there was one occasion where I flew an arrival into Bangkok from Frankfurt on the 747 with a hydraulic system failure. This involved a flap and landing gear extension using a backup system. With good planning, this was not usually a drama. Because the flaps took about six minutes to run out using the alternate system, the flap extension would be started at about 30 nm out, with the gear extended at 15 nm out. But this time, Bob was nowhere to be seen and things didn't go quite to plan. When the alternate landing gear extension was complete, the system showed the wheels were not locked down. The best we could hope for now was that the gear would lock down on touchdown, or at least not collapse. The worst thing would be a gear collapse on landing.

There wasn't much time to muck around, with just over an hour of fuel remaining, so we declared an emergency and asked for a heading away from the field to prepare for an emergency landing. We got the preparation started in the cabin and were in the process of briefing for the emergency landing in the cockpit when we went into another turn. In this turn, the g-forces, or the angle of bank, finally made the landing gear lock down. After a successful landing, it was time to reflect. For a few days after that, I was a bit shattered by the stress of having thought I was going to possibly land wheels-up in Bangkok. Imagine the paperwork.

FLIGHT CONTROLS AND COMPUTERS

Orville Wright only had one flight control computer and that was in his noggin. An Airbus A380 has seven: three primary computers, three secondary computers and a backup control module. Because modern flight control computers send signals to other systems to operate the flight controls, pilots receive no direct feedback from the flight controls. This means artificial feedback is required, and the computers must be programmed to not over-stress the aircraft with overly rapid control movements. Failures of flight control computers are rare and there are enough backup systems to cope.

Sometimes these highly complex systems can be an issue, especially if they remove the pilot's power to have the final say. The book *No Man's Land*, by ex-Qantas Captain Kevin Sullivan, is about such an event. Here's a short summary of it written for me by Captain Kev, himself: *QANTAS flight QF72, an Airbus A330, departed Singapore on 7 October 2008. It was a scheduled flight with 315 passengers and crew aboard. The weather en route was fine and the aircraft was completely operational.*

The A330 is controlled by a computerised fly-by-wire system; there is no mechanical link between the pilot's control stick and the flight control surfaces on the wing and tail. Additionally, the aircraft's computers are programmed to maintain the aircraft within a safe flight envelope at all times, thereby enhancing safety through-out its operational regime. It is the primary flight control computers that move the control surfaces electronically in accordance with the pilot's commands.

As the aircraft approached the northwest coastline of Australia, in clear skies and an altitude of 37,000 ft, the aircraft's computer systems began issuing warnings of impending 'STALL' and 'OVER-SPEED', essentially warning the pilots that the aircraft was operating at its minimum and maximum speeds simultaneously. The auto-pilot disconnected and the captain was flying the aircraft manually. Without warning, the aircraft's nose pitched down abruptly and violently without pilot input, catapulting unrestrained passengers and crew into the ceiling of the aircraft cabin. After two uncommanded pitch downs, 119 passengers and crew were injured, many seriously.

A 'MAYDAY' was issued by the captain and the aircraft diverted for an emergency landing at Learmonth airfield near the town of Exmouth, Western Australia.

Despite the designed redundancy of five flight control computers and associated protection modes to keep the aircraft safe, a 'ghost in the code' in the form of a design limitation allowed the aircraft's air data computer to generate extreme and erroneous values of angle-of-attack, and the flight control computers to act on the data as if the aircraft was about to stall. It was not. The pilots were powerless to prevent the pitch downs as their control inputs were blocked by the flight control computers.

Complex systems create complex failures. QF72 was a science fiction scenario that became a reality; computerised systems designed to keep the aircraft safe were in fact trying to harm all onboard. It was unprecedented and unforeseen. Fortunately, the three pilots were able to control and eventually land the aircraft despite the loss of many essential systems.

There is no doubt that increased automation in aviation has enhanced safety and reduced accidents, but it has introduced complex failures that may not be covered by the manufacturer's operating manuals. The three pilots were on the verge of being over-whelmed by the incessant aircraft warnings and cascading failures and faults of essential systems caused by confused computers reacting to an erroneous, self-generated situation.

In recent years, the argument to remove one pilot from the cockpit during the cruise phase of a flight is based on the increased advancement of automated aircraft. This is folly. Pilots are still in the cockpit to save the day, and it is easy to see that a single pilot on duty during failures such as experienced on QF72 could have produced a much different outcome.

In this battle of pilot versus computer, the pilots prevailed and delivered their passengers to safety.

AIR AND PRESSURISATION

I've covered this previously in the book, and there's not much more to say about the systems. But a sudden loss of pressure in the cabin at high altitude, known as a rapid decompression, can be quite frightening. Although it's rare and I never saw one, it does happen from time to time. Pilots are frequently trained for this in the simulator.

The first sign of it will likely be a loud bang followed by a sudden fogging of the cabin air. The oxygen masks should deploy all around you, and you should get one on quick-smart or you will pass out thanks to hypoxia. If hypoxia doesn't get you, the smell from high altitude flatus evacuation (HAFE) will. In a rapid depressurisation, the process you endured slowly during the climb, of expanding air in the body, is suddenly accelerated well beyond a comfortable level. In the galleys, thanks to the reduced cabin pressure, water will now boil at 63°C at 33,000 ft.

On the flight deck, the pilots will, after initially weeing themselves a little, put their own masks on and commence a rapid descent to bring the aircraft down to a level where supplementary oxygen isn't required. In the case of Qantas Flight 30 on 25 July 2008, this only took five minutes. QF30, a Boeing 747-438 aircraft carrying 369 passengers and crew, rapidly depressurised following the forceful rupture of one of the aircraft's emergency oxygen cylinders in the forward cargo hold. The aircraft was cruising at 29,000 ft and was 55 minutes into a flight between Hong Kong and Melbourne. Luckily no one was injured in this event.

COCKPIT FLIGHT DISPLAYS AND THE AUTOPILOT

All the cockpit instrumentation and the computers that supply the information are critical for the safe and efficient operation of the flight. There are multiple backups for both the equipment and the sources of incoming information, so any one failure on its own usually isn't very significant.

The primary flight display is a critical instrument for pilots, because it displays the attitude, or orientation, of the aircraft relative to the horizon. In cloud, without it, the pilots would have no idea which way was up or down. If a pilot's attitude indicator fails while flying in a cloud, it can have severe consequences. There are two other attitude indicators on each flight deck, and if one fails, pilots are trained to check all three to decide which one is faulty and which ones to use. This type of failure is frequently trained for the simulator, and it takes a lot of self-discipline to do it right.

I had an attitude indication failure happen to me once on approach to Bangkok. Bangkok, at certain times of the year, is known for its thunderstorms. On this particular day, as we approached the airport from the north, a storm had moved onto the final approach at the 12 nm point. We asked air traffic control for a diversion around it with the hope of recapturing the approach path closer in, but as I turned to the right in cloud and heavy rain, my primary attitude indicator failed, and I had to crosscheck the other indicators. The other indicators confirmed mine was the failed instrument, so I had to revert to using the standby indicator.

The fact that this happened in those conditions, just after I had entered a turn, shows how, sometimes, 'Murphy' is following you. An incorrect reaction from a lack of training could have put that

747 completely on its back at 3,000 ft above the ground in cloud. All those simulator sessions really are worth something.

The primary flight display also shows the aircraft's airspeed, and this is critical as well, so pilots are trained over and over to deal with errors in the airspeed indicating system. In the crash of Air France 447 in the middle of the Atlantic Ocean in 2009, air speed errors were to blame. The investigation concluded that the aircraft suffered temporary inconsistencies between the air speed measurements – likely resulting from ice crystals obstructing the aircraft's pitot tubes, which caused the autopilot to discon-nect. The crew incorrectly reacted to the abnormality, causing the aircraft to enter an aerodynamic stall from which it did not recover. As a result, training for this problem is now at the top of the list.

Most of the time, the autopilot is your friend when the going gets tough in the cockpit. An autopilot is a computer linked to the flight controls that flies the aircraft for the pilot. The pilot must still tell it what to do by pushing the right buttons and setting up the correct parameters, but once the autopilot has been told, it will fly the aircraft on its own without further manual input from the pilot. If it's working and not affected by other problems, then it takes a load off your mind.

WHEELS AND BRAKING

The Wright brothers had neither wheels nor braking, so the pilot either landed on skids or on his face, depending how the flight went. In the early days, when wheels were first added to planes, they were fixed in place. The downside of this was drag, so even-tually engineers designed retractable landing gear that could be raised and hidden from the airflow. These retraction systems are complicated and heavy, so it's a trade-off. On modern jets, the economics of the reduced drag outweigh the increased complexity and weight of the landing gear systems.

Of course, having the wheels up is only worthwhile if they will come down again in time for landing. Over the years there have been many stories of aircraft landing with their wheels still up. This is normally because of forgetfulness or something breaking, and if you forget, something normally breaks, so there's a lot of interdependencies going on here.

In one famous story, an aircraft was on approach for landing. The observant tower controller noticed the wheels weren't down, so he instructed the pilot to carry out a missed approach and not land. But the pilot continued and landed with no wheels. Later when asked why he didn't follow the controller's instructions, he said that he didn't hear them because of the loud horn going off in the cockpit. It was the landing gear warning horn.

These days, aircraft wheels and particularly tyres are quite impressive. Aircraft tyres can run at pressures of around 220 psi, but are built for up to four times that pressure before they will explode. Each large aircraft tyre can support close to 30 t of weight on its own. And they have to be tough, too, because on landing, the g-forces and sideways forces if the pilots stuff it up can be huge. But despite that, it's rare these days for a tyre to explode. If one does explode, you don't want to be near it when it happens. It will definitely hurt the next morning. People have been killed by exploding aircraft tyres.

Tyres are filled with inert nitrogen so that they won't explode under conditions of high heat, but they can still shit themselves, and the damage done by large chunks of rubber travelling at high speed can be significant. The nitrogen also serves the purpose of preventing rust inside the wheel and reducing expansion due to temperature changes.

I had a 747 tyre explode 20 years ago on take-off from Bangkok. We didn't know about it until the tower controller advised us that they had found rubber on the runway after take-off. Passengers in the cabin had reported hearing a loud bang and claimed that they could still hear an unusual wind noise. The second officer went to the cabin and confirmed the noise.

We chose to fly onto Sydney and carried out an emergency landing on the remaining wheels with fire trucks at the ready. It all went pretty well, and after arriving at the terminal I went down under the aircraft to have a look. There was a hole in the side of the aircraft near the blown wheel around half a square metre in size. These days, modern aircraft have tyre pressure reading systems, so you could confirm something like this without even leaving the flight deck.

Aircraft tyres also have fusible plugs on the inside of the wheels, which are designed to melt at a certain temperature. Tyres often overheat if maximum braking is applied during an aborted take-off or an emergency landing. The fuses provide a safer failure mode. They prevent tyre explosions by deflating in a controlled manner, thus minimising damage to aircraft and objects in the vicinity. In a heavy-weight, high-speed rejected take-off, it is not unusual for all the main wheel tyres to deflate in this way.

Depending on your source, the tyres on a large aircraft have to be replaced after every landing (Adelaide taxi driver) every 100 landings (a ground engineer I knew) or 300 landings (the Internet). The reality is that it varies depending on the number of ground miles taxied, the type of surface on the taxiway, how heavy the aircraft is, how fast the aircraft is taxied at, how many corners they go around on 11 wheels (an A380 has 22 wheels in total) and how rough the landings are.

Tyres on the larger aircraft retail for around US$5,000 each, so with an A380 having 22, I'd be doing a ring-around before committing to a new set. And I'd definitely be demanding a wheel alignment be thrown in.

Being able to stop an aircraft when needed is something pilots are reasonably enthused about, so braking systems are pretty important. Modern aircraft have multiple backups for braking, plus spoilers and reverse thrust if needed. Spoilers are the metal plates that rise out of the top of the wing from time to time. In flight they are used to reduce lift and increase drag to increase the rate of descent. On the ground they kill all lift and increase drag to help slow the aircraft down.

The braking systems include automatic brakes, which are programmed to operate at a chosen intensity straight after landing. The latest cool feature is 'brake to vacate', which allows the pilots to pre-program the automatic brakes to bring the aircraft to a taxi speed at a chosen exit onto the taxiway. Anti-skid comes standard on all models of aircraft as well.

FUEL SYSTEM

The object of the fuel system is to get fuel from the tanks to the engines. This gets quite complicated on large aircraft. The A380 has 11 separate fuel tanks, 10 in the wings and one at the rear that helps keep the aircraft balanced. These modern fuel systems are able to automatically move fuel around between the tanks as required to maintain balance. There's not a lot that can go wrong because if certain fuel pumps fail, other pumps can provide pressure as a backup. Most aircraft engines can also operate from fuel being fed by gravity alone.

If a leak is detected in any tank, that tank can be isolated so that all of the fuel onboard doesn't run out through the leak. Once this procedure is carried out, there should still be enough fuel in the other tanks to make a suitable airport. If not, then it's time for some boating. Over water, a ditching would be required. A ditching is not an ideal scenario, especially if you are flying in polar regions or over stormy seas.

Given that the fuel is not going to run out instantly if there is a leak, there are actions a crew would initially take. Aircraft over cold regions would head towards the equator so that the ditching could be carried out in warmer water. It's worth noting that the majority of sharks like warm water too, just saying. Air traffic control would also advise of the position of any nearby ships, if you asked nicely. Plonking an aircraft down in the water next to a ship would be ideal for a quickish rescue.

FIRE PROTECTION

Fire on an aircraft is probably the worst thing that can happen, which is why all crew are extremely tense about it. The problem with fire is that you can't run away from it on an aircraft. Because of this, there are fire protection systems for engines, cargo compartments, toilets, the inflight entertainment and other critical areas. All crew are trained annually in firefighting techniques as well.

If a fire is not immediately brought under control, then other action needs to be taken post-haste. Apart from the odd toilet fire-extinguisher going off, it never happened to me, but if it had, my plan A was to descend quickly to a very low altitude. Then, if the fire was brought under control, I could relax a bit and make a new plan to divert to the nearest suitable airport. If the fire was still out of control, then an emergency landing or ditching anywhere would be my preference rather than getting barbecued.

COMMUNICATIONS AND THE INTERNET

The Wright brothers used voice for their communications during their hundreds of practice glides and flights. 'Where did you get your pilot's licence, Wilbur? The Weet-Bix packet? That was bloody awful.'

'Fuck off, Orville. You do better.'

I mentioned previously how the high frequency radio was awful to use. It was sometimes so bad that winding down the window and yelling to Bombay seemed like it might work better. Sadly, that wasn't an option, but only because you can't wind down the window. Airlines have to pay more for that option. Happily, these days, HF radio is nearly a thing of the past.

These days very high frequency (VHF) radio covers vast countries through the use of repeater stations, and modern aircraft now have satellite communications and an Aircraft Communications Addressing and Reporting System (ACARS). ACARS is a digital data-link system for transmission of short messages between aircraft and ground stations via VHF radio or satellite.

For Internet, aircraft use both ground and satellite systems. The ground system is perhaps the more basic of the two. Essentially, an aircraft is tuned into a wi-fi hotspot, with an antenna fitted at the bottom of the fuselage. As the plane cruises through the sky, the antenna searches for the most viable transmitter, disconnecting and reattaching to give the best possible connection onboard.

The only issue with this method of Internet delivery is that it doesn't work over water. Crossing the Atlantic Ocean, for example, passengers would be disconnected from streaming and messaging services for a large portion of their journey, and their social media lives would cease to exist for longer than is comfortable. For that reason, airlines also use satellites. To connect to orbiting satellites, aircraft must be fitted with an antenna on the top of the plane. This antenna connects with the closest satellite, which passes information between the aircraft and the ground. The router on the aircraft distributes the signal throughout the cabin.

OTHER EMERGENCIES, FAULTS AND FAILURES

Given the number of years I flew big jets, the incidences of systems failing was very small. I did have an interesting one occur on a flight to Santiago in Chile on a 747 a few years ago. We were flying along, minding our own business during the daytime, when suddenly my windscreen cracked. It was a big multi-pronged, poo-inducing crack. My first reaction was to shit myself, because my brain was telling me that the whole windscreen was going to implode on me, and I would soon be flying and dining al fresco into a 900 kph breeze.

But the windscreen on a modern jet is triple layered and about an inch thick. Usually, they fail because moisture ingress eventually causes delamination, heat-coating problems and arcing, which, if not corrected, lead to cracking of the outer ply. My crack was also only on the outer ply. This is not that uncommon, so, given that a replacement windscreen could cost around US$50,000, airlines would be wise to take out windscreen cover on their insurance policies.

So, there we were, about four hours east of Sydney with a cracked windscreen, and being the compliant captain that I was, I left it up to the company to decide our next course of action. With this approach, you run the risk of getting bad advice, but it also means you can't be blamed for doing the wrong thing. In this case it seems they must have asked a baggage handler for his opinion, because their answer was for us to continue to Santiago and get it fixed there.

The problem with replacing windscreens on aircraft is that you can't just go to Windscreens O'Brien and pick up a 747 windscreen. There were none available in Chile. Once they finally got one to Chile, no one there knew how to replace it – and there was the added complication that you can't reach the window standing on tip toes. When you do finally find an engineer who is 15 m tall, he will take four hours to change it, then the glue has to be given quite a few hours to dry as well. As a result, we had a four-day stopover in Santiago with three other crews. If I trace it back, that's where I believe my liver problems first began.

We've now covered most of what can go wrong. Engine failures are something all pilots train a lot for. The most critical time for that to happen would be on take-off after passing the point of no return. On a four-engine aircraft, you'd lose only 25 per cent of your available power if you lost an engine. On a modern twin jet, you'd lose half of your power, but the amazing thing is that the remaining engine can still cope.

Engine failures are no longer common because of the new technologies involved. The majority of engine failures are simple failures for a number of reasons, like lack of fuel or airflow disruption. Often a restart can be attempted immediately. It's rare to get an engine failure that involves severe damage and bits of engine flying everywhere, but they do still happen.

The important thing is for pilots to be well trained in how to handle any emergency. Once the initial emergency actions have been taken, then the autopilot should be used to unload the pilots so that checklist actions can be completed slowly and methodically. The number one priority is to continue to fly the aircraft and not be distracted from that.

EMERGENCY LANDINGS

The end result of dealing with a major issue can sometimes be that the checklist tells you to land ASAP, so we need to talk about emergency landings. If the issue is not serious, such as if one of four engines fails, the checklist might say you can trundle on for a bit. The failure of an engine on a two-engine aircraft is more serious, because you only have one engine left. So in many cases you need to get your wounded bird onto the ground.

This generally means a diversion to the nearest suitable airport. 'Suitable' means the aircraft can legally and comfortably land there, the weather there is okay, and there are adequate services to assist you when you do land.

Getting to the ground safely is the number one goal. Sometimes this can be complicated by inoperative systems or damage, but generally there are procedures to follow which cope with all this and allow you, with a bit of thought and careful planning, to get it done.

Of course, even getting to the ground in one piece is no guarantee of survival. When an Asiana B777 crashed at San Francisco

Airport in 2013, one of the three people killed was run over by a rescue vehicle.

Once on the ground, it may be possible, under the supervision of emergency services, to taxi the aircraft to the bay for a normal disembarkation, but if there is a lot of damage or a fire, it will be necessary to evacuate the aircraft.

During certification, aircraft manufacturers are required to demonstrate that their aircraft can be completely evacuated in 90 seconds in an emergency. So, the manufacturer pays a plane-load of 'passengers' of different ages, genders and sizes to be guinea pigs. The manufacturer then blocks a couple of exits, introduces smoke into the cabin and sets off the alarms to prove it can be done. This is from a stationary, unbroken aircraft. The reality may be much worse, with cabin bags and injured people blocking the way out. In the pre-flight emergency briefing, passengers are always advised to leave cabin baggage behind, but many won't hear the instruction, and others won't think it applies to them.

At over-wing exits in smaller aircraft such as the A320 or B737, where passengers are required to remove the windows so that people can get out, you are already relying on quite a lot of luck for it to go smoothly. These over-wing doors, in my opinion, are a last resort. I would personally be running away from them. This is because they are more like a window in size. So instead of walking out a door then jumping down a slide, you have to squeeze between some seats and somehow get through a window-sized hole. If a large person is stuck in the door like Winnie the Pooh, this will make it a lot harder.

If you succeed in getting out onto the wing and you turn the right way on an A320, there will be an automatically inflated off-wing slide to zoom down. If you are on a B737, there isn't a slide, and the drop onto the solid concrete is around 3 metres. Part of the evacuation checklist for the 737 pilots is to lower the flaps so people can slide down them, but this won't slow you down. I don't know about you, but I would probably break something jumping 3 metres onto concrete. All is not lost, though, because if you wait for a bit, a pile of broken people will build up, and eventually you will be able to step off onto the top of the pile. Patience is a virtue if your arse is not actually on fire.

If an aircraft runs into the water during take-off or after an emergency or bad landing, the evacuation becomes a ditching procedure. It doesn't hurt as much leaving the aircraft, but the possibility of drowning arises. It pays to listen when they tell you where your lifejacket is stored in the safety briefing.

If a damaged aircraft is unable to make it to a suitable airport and is over water, a planned ditching will need to be carried out. In modern times these are as rare as hens teeth, which means we must be due for one.

Ditching an aircraft wouldn't be fun, even if it was on the Hudson River with Sully. Your best plan is to leave it to the more adventurous passengers. I used to laugh at how the passenger emergency briefing sheets used to say, 'When the aircraft alights on the water . . .' It sounds like a frog doing ballet on a water lily. It wouldn't be like that, especially in rough seas. There could potentially be a lot of damage to the aircraft, and even if you do subsequently make it into a life raft in one piece, you will soon wish you were dead. I've spent time training in life rafts in a swimming pool, and I got seasick within a few minutes. And imagine if you were squeezed in a raft like sardines next to someone who was boring and never shut up. A swim with sharks suddenly looks very attractive.

CRM

One of the most important things to happen in aviation in the last 40 years is the realisation that good CRM, or crew resource management, really enhances safety. It is especially useful when stress levels are high in dealing with emergencies. CRM is a set of procedures for use in environments where human error can have devastating effects. It is primarily used for improving aviation safety, and focuses on interpersonal communication, leadership and decision-making in aircraft cockpits.

Once modern aircraft began to use flight data recorders and cockpit voice recorders to monitor accidents, it was discovered that these accidents often occurred where crews did not respond appropriately to emergency situations. Crew members might not communicate clearly or work as a team. CRM was developed to help guide emergency decision-making.

In English, CRM means the captain, instead of being a godlike one-man band, accepts that they are human and seeks input from the other crew members, because three heads are better than one. CRM training also trains the crew to not be like a bunch of sheep and simply follow what everyone else is thinking. This inclusive-ness also extends to working with the cabin crew when required, as the whole operation needs to be as efficient as possible.

Towards the end of my second year in Anchorage with Japan Airlines in the 1990s, JAL suddenly got a bee in their bonnet about CRM. The Asian airlines were known, at this point, for their 'steep cockpit gradients', where the captain was God and no one would question him. This had led to quite a few accidents over the years. Good CRM involves using all the skills available on the flight deck, and to achieve that, crew members must be encouraged to speak up when they aren't happy.

Because JAL was acknowledging how bad they were at this, they decided we foreigners, who were traditionally much better at it, needed to attend an expensive week-long course run by Scientific Methods in Houston, Texas.

On the first day of the course, we met Jim, who was running it. Jim was an ex-United Airlines captain. So as not to bore you, we spent a week learning that you can either be people-oriented or task-oriented. Some people can be both. It was scaled from 1 to 9 on a graph. If you were the sort who cared about people and not about the task, you'd be described as 1 for task and 9 for people, or 1,9. If you cared only about the mission and not at all about the people, you'd be 9,1. If somehow you were the ultimate CRM hero, you'd be a 9,9. This is the holy grail and I doubt it is possible.

The CRM course went on for five days. I suspect that this was to justify the cost, as it could have been covered in two. It got very tedious and repetitive. They even put us in rooms after dinner for a few hours for group discussions, and Jim would pop in from time to time to see how we were going. We took to locking the door, so when he knocked, we could turn the TV off, and who-ever's turn it was would jump up and pretend to be pontificating about something.

After about the third day, we were well and truly over it. Some of us had brain damage from boredom, but Jim would not let up. Poor old Jim was very 9,1. We all survived the week, and all in all I learned a lot from the course.

One thing that I do remember coming out of the course was that, even with CRM, the cockpit is not a democracy. It just doesn't work that way. At one point during the course, they put us in groups that had one captain and a few first officers and engineers. Then they gave us a problem to solve. No one seemed to be taking the lead and cocktail hour was fast approaching, so I dived in and said, 'Right. Let's gather all the information we need, decide on a course of action and then I'll dish out the tasks.' The young engineer in the group asked, 'Why can't I hand out the tasks?'

I said, 'Because you aren't the fucking captain, mate.' Everyone else nodded sagely in agreement.

Shortly after we got back from Houston, I had to go down for my six-monthly simulator check in Tokyo. The JAL training depart-ment added an extra session to this check for CRM training. For this, we got airborne out of Haneda with about 120 minutes of fuel, and from the moment we got airborne, things started to fail. I spent the next hour ordering checklists one after another while I prepared to land back at Haneda. More things kept failing all the time. I can't remember how many different emergency checklists there were on the B747, but if there were 73, we did them all and five of those twice. Eventually, I threw it on the ground and taxied in, with more things broken than working. In the debrief, the JAL instructor said, 'Ahhhh, Captain Burfoot-san. Not much use of CRM.'

I said, 'There wasn't much time to hold a committee meeting. We were just trying to survive.'

He looked confused. I hope they have gotten better.

I won't deny that all pilots benefited from good CRM training, including those of us who already thought we were bloody fantastic. Actually, especially those pilots. And the reference to the Asian problems of severe cockpit gradients is a fair generalisation, but bad CRM techniques keep popping up in all airlines. The Air France 447 accident, although blamed on erroneous air speed readouts, was also a famous example of piss-poor CRM.

Prior to the Scientific Methods CRM course, and during my initial training with Japan Airlines, I did a training trip with a grumpy German called Jurgen Zoch. Zoch was a very unpleasant man. I suspect that he may have been a teenager around 1945 in Germany, and this would have had a profound effect on anybody, so later, I cut him some slack. But back to the flight. Initially, we got airborne, heading for Frankfurt off runway 34 at Narita with me flying. By 500 ft, Zoch was yelling at me even though I was flying perfectly. This went on for a minute or so, until I decided that this wasn't really the place for arguing. I turned to Zoch and said, 'Either shut up and let me fly or you can take over.' So, he shut up. Later there was no explanation or discussion, so I let it go. After that, he was a different person all the way to Frankfurt and back, and quite friendly when I ran into him after that. Zoch was to CRM what Scrooge was to Christmas.

NAVIGATION

Unkind people who know me will tell you that the reason I never get lost is because people are always telling me where to go, but the real reason is that I started my aviation career as a navigator in the Royal New Zealand Air Force. One of the most important things I learned there was that you can get a good idea of direction from looking at your watch and the position of the sun. In London, at around midday, for example, south will be pretty much in the same direction as the sun. If you are lucky enough to have a good sense of direction, be grateful.

When I started as a navigator in 1978, I was using a drift sight, automatic direction finding (which was neither automatic nor capable of finding a direction accurately), sun and star shots and the Mach I eyeball. Later when I graduated onto the P-3B Orion, I was spoiled with Loran, or long-range navigation; Omega, a single, early inertial navigation system (INS); distance measuring equipment (DME) and radar fixes. Only two of those options are still in use today, DME and the INS.

INS has come a long way since I first used the single system on the P-3Bs. It was an early system based on a platform with accelerometers and gyroscopes, and its calculated position could easily drift 15 km per hour from the true position. We navigators used to work out an error vector for the INS that we would update using good fixes. It was a known joke to tell the crew that although we were heading for Hawaii, it seemed the INS preferred Tahiti and was heading that way now.

The technology of INS navigation has improved since then, but the concept is the same. The outputs from accelerometers and gyros are integrated using integral calculus to give speed in any direction, which in turn is integrated to provide distance moved, which logically follows with a position and altitude update.

By the time I joined Qantas, the 747s had updated INS versions, and there were three of them comparing their positions with each other. Picture three hikers arguing about which way to go and all heading off on their own in frustration. We pilots were able to 'drag' the errors back in by updating them from distance measuring equipment range circles, much like a ranger would round up the three lost hikers.

Later updates of INS removed the need for a platform and moving parts, which got rid of friction errors. These updated systems are called IRS, or inertial reference systems. In an IRS, the accelerometers never actually move. They use electrical current to stop movement, hence the current required is equivalent to the acceleration. The gyroscopes in the IRS are now laser instead of spinning.

If you are technically minded, this is for you. Laser gyros are still subject to minuscule errors from 'injection locking'. The fix for this is 'forced dithering'. Feel free to do your own research because this sort of detail is only on a 'want to know' basis, but personally, every time I see someone dithering now, I think to blame injection locking.

Suffice to say, IRS/INS these days are super accurate and reliable, and are backed up with triple global positioning systems (GPS) like the ones in your phone and car. The combined accuracy of these systems can be within a couple of feet. It's hard to get lost unless you are a goose or your name is Dufus, or both.

You might be wondering why, with modern GPS available, we still use IRS navigation. There are two reasons. Firstly, they provide other information to instruments systems such as aircraft attitude and track, and secondly, they are self-contained on the aircraft, so even if baddies jam up the GPS systems, the IRS will still be your faithful friend.

Because aircraft fly over long distances at high speed, they fly on great circle routes. Great circles are the shortest distance between two points on the Earth. A great circle is a line that curves away from the equator on a standard flat chart. The reason for this is that unless you are using a globe, it's impossible to depict a curved surface on a flat chart. If you put a piece of string between two points on a globe, that will be a great circle. So when you see a curved line depicting your route on the inflight map, you don't have to wonder why anymore.

Anyway, I'm sure some of you have glazed over and are sucking your thumb right now, and I apologise for that, but navigation is important. It's improved out of sight from when I started my career, and that's a good thing. The accuracy and enhanced displays of navigation data on the flight deck have added a substantial layer of safety to aviation.

Now buckle up, because shit is about to get real. It's time to prepare for descent and landing.

SIX

LANDING

RETURNING TO EARTH WITH A THUMP.

PREPARATION FOR DESCENT

On a longer flight, this preparation begins a few hours out from the destination. Many international flights are overnight flights, so this preparation will start with you passengers being woken just minutes after you finally fall asleep.

If you are flying on a premium airline, you might even be offered breakfast, even though this is the last thing you could possibly want. My advice is to eat what you can and stick the other poached egg in your pocket for later. A few hours down the track, when you don't have time to eat, you will wish you had done so.

Other things you need to take care of is your '0.4' visit to the loo, and quickly, because everyone else will have the same idea. If you haven't filled out customs documents on an international flight, now is the time. You also need a plan for what you will do on arrival. The reasons for a plan are twofold; A: Because everyone needs a plan. B: Because your tired and possibly jet lagged arse won't be operating at 100 per cent in an environment where less than 100 per cent is not ideal – that is, an airport.

In business and first class, life is even more stressful, because passengers there need to remember to change out of their pyjamas. I had a nightmare once about not doing that and woke up screaming. My fears were confirmed when I realised I was still in my pyjamas.

Meanwhile, in the cockpit, everyone is back from the crew rest, and they are hunting down a nose-bag to throw on and some coffee to slam. The next priority after that is to get ready for the arrival. This is a big deal. When you are hurtling towards a full stop at 950 kph, things start to happen very quickly, so pilots like to get everything that they can out of the way in advance.

This involves setting up the computers for the expected arrival based on the weather at the destination, which you would have recently updated. Most crew will also look at and load any requirements for a last-minute diversion to an alternate airport. Burfoot's fourth law states that the more preparation you do, the less likely it is that something unexpected will happen.

And then it's time for the PA to the passengers. Some pilots will waffle on for ages with this, filling people in with details of time zone changes, who won the rugby, if it's anyone's birthday, and the fact that they have a choice of airlines (why would you tell anyone that?). People really only want to know if it's raining and what time they are getting there. Everything else is superfluous.

About 20 years ago, someone in the change department decided it would be a good idea for pilots to make PAs in front of a flight-deck camera, which would then transmit the video to the inflight entertainment screens. But someone in the department dropped the ball and decided to consult with pilots first. I sent them a response asking them if they had seen what I looked like at 5 am after being up all night. The idea was quietly shelved.

Mostly the preparation for descent is fairly generic, but for certain airports in the world, there's a lot more to it, because of bad airport design or challenging terrain near the airport. The old Hong Kong international airport, Kai Tak – which operated from 1925 to 1998 – was one of these. The airport's runway was bordered by water on three sides, and Kowloon City's residential apartment complexes and 2,000 ft plus mountains to the northeast. This made it impractical for aircraft to fly over the mountains and drop in for their final approach. Instead, planes flew over Victoria Harbour, Kowloon City and Bishop Hill in the north. Pilots would search for Checkerboard Hill, which featured a large red and white checkerboard pattern, then make a sharp right turn at low altitude to start the short final approach and touchdown. The approach was a challenge on a good day and a real nightmare when the weather was terrible.

The typhoon season in Hong Kong lasts from May to October, and the weather can get very marginal. On approach, you could be in the cloud until the last moment, and then you'd only have a split second to get a good visual reference before starting the turn, possibly with rain on the windshield as well. With the weight and inertia of the big jets, you also had to be aware of which direction the wind was coming from. If there was a crosswind on the runway itself, you couldn't turn the corner and line up the nose with the runway, as you'd be blown off course with not much time to correct your error. Instead, you had to anticipate and finish the turn already offset to account for the drift.

If this is all too technical and you are starting to stretch and yawn, spare a thought for the 'average pilots' from some of the lesser-known airlines, who were massively stretched by the requirements for this airport. Many had engines strike the ground before the wheels, and a few ran off the runway into the sea. It was considered outstanding entertainment to go there and find a bar overlooking the airport on a windy day to watch the entertainment when things heated up for the pilots.

The final 47-degree turn was also interesting, because the aircraft passed level with and between apartment buildings. If you were brave enough to take a quick look, you could see right into people's living rooms as you flew in. Urban legend has it that one particular Chinese man used to 'moon' all the aircraft as they passed by. I never saw him, but it sounds likely, based on human nature.

One day, out of the blue, the Hong Kong Government decided to build another airport called Chek Lap Kok. It would be built on reclaimed land on the northwestern side of Lantau Island and include Chek Lap Kok Island. Actually, it wasn't entirely out of the blue. In fact, by Hong Kong standards, a fair amount of dithering went on, not helped by interference from the Chinese Government in Beijing. But once construction of the airport and all the supporting infrastructure of accommodation, roads, bridges and train services commenced, it only took them about five minutes to get it done. Are you raising your eyebrows? Well, five minutes might be a stretch, but you know what I mean. I'm sure they didn't spend too much time consulting with the local indigenous people or bird watchers and environmentalists. Neither would they have employed an army of safety consultants wearing high-vis vests and carrying clipboards and concerned looks.

In the end, the only person who didn't win from the construction of the new airport was the Chinese bloke who used to moon the aircraft. I hope he's okay.

DESCENT

And just when you were thinking the flight would never end, the engines suddenly go almost silent and the nose of the aircraft subtly drops, signalling the start of the descent. The pilot flying, who will have worked out the ideal point to start the descent, will now be thinking, 'Bring it on, air traffic control. Whatever you try, I'm up for the challenge.'

At the start of a normal descent from, say, 37,000 ft, you should be about 130 nm or 200 km from your destination. Since descents are made at idle thrust, that's a pretty impressive glide for a large heavy-weight aircraft. You'll still be whistling along at 950 kph too. Somewhere in between this point and the landing, the pilot needs to reduce to approach speed of around 250 kph and be at the right height to start the approach 20 km out. The way to do this is to have speed and altitude targets at certain points and heights to check how the whole energy management scenario is going.

When pilots first start flying in a small training aircraft, it trundles along at around 105 kt. As their careers progress, the aircraft get bigger and faster. I've known pilots who have struggled with the increase in speed and have failed to progress. They usually say, 'Everything is happening too quickly for me. I'm used to 2 miles a minute, not 3 or 10.' Good pilots don't think this way. They think in terms of time. 'How long do I have until landing? Thirty minutes.' That can be 50 nm or 200 nm, but it's still 30 minutes to get all the same stuff done. Speed is a mindset problem.

But getting back to the air traffic control challenge, they can really screw you around. They can make you level out and get high. They can turn you off track on a vector (heading) so the extra track miles will make you low on your profile. If you get high, you can use the speed brake which raises the spoilers on the wing, dumps lift and increases drag. But this move is not popular with passengers, because it makes the whole aircraft shake. On the 747, using the speed brake was an admission of defeat. On the A380, it was more common due to the aerodynamic characteristics of the wing and not as shameful to use it.

Just because you are on descent and the pilot is sparring with air traffic control doesn't mean that it's all over bar the shouting. If the weather at the destination is marginal, there's a slim chance you may end up on some sort of mystery tour at no extra charge. On one particular flight from London to Singapore, we had such an incident. At the start of the descent into Singapore, the forecast indicated thunderstorms around the field. When thunderstorms are forecast, we legally require fuel for an approach, then a diversion to an alternate. We didn't have this, but we continued towards Singapore as long as we could in the hopes that the weather would improve.

It didn't, so partway down the descent, we requested an inflight diversion back to Kuala Lumpur. The Singapore arrivals controller started to argue that aircraft were still landing in Singapore, and he did nothing about it. So, we asked for a diversion again. There was still no response. By now, we were right on minimum fuel for returning to Kuala Lumpur, so I said to the first officer, 'Squawk 7700 and declare an emergency.' Which he did. Squawking 7700 on our transponder would be something they couldn't ignore. Air traffic control would instantly get an emergency indication show up on their radar screens. We got immediate action from that and ended up safely on the ground 30 minutes later in KL.

Some hours later, when the weather had cleared in Singapore, we landed back there. I was then phoned by the duty controller, who told me that I had caused a significant drama, because when 7700 is squawked, all the emergency services and local hospitals are automatically contacted. I pointed out to him that perhaps he should ask his people why it had happened at all. If they had have given us what we asked for in the first place, it wouldn't have.

If a lot of aircraft are arriving at an airport at the same time, air traffic control has to delay them all. To do this they can vector aircraft off track for a while or make aircraft enter a holding pattern. New York's Kennedy Airport was always good for vectors. Coming in from the west, they'd fly you right over the top of Manhattan then send you down the coast to the south. This got a bit silly at times, as there'd be a line of aircraft extending almost down to Atlantic City, all burning a heap of fuel at low level and all with thirsty pilots on board. I swear one time we went so far south before turning back to the airport that I could see the Florida Keys.

One of the worst things about vectors and lines of aircraft is that in some countries the controllers are too patient with flights from countries where flying aircraft is still seen as hard work.

I won't say which countries, or I'll risk getting a truckload of old rice dumped on my front lawn, but certain airlines are very good at holding up the works by slowing down 40 nm from the airport and coming in at minimum speed so the pilots can cope. This is very inefficient, because everyone behind them has to slow down too. As a pilot, there's nothing you can do about it except get out the cards for a few hands of poker.

Holding patterns are the other delaying tactic used by air traffic control. These are big oval or 'racetrack' shapes to trace in the sky, usually lined up overhead a navigation aid. They can take six to eight minutes to go around once. London Heathrow is a great place to experience holding patterns. Heathrow deals with around 650 arrivals per day on average. There are four holding patterns or 'stacks' at Heathrow, known as Bovingdon, Lambourne, Ockham and Biggin.

Depending on which direction you are flying in from, at busy periods you will end up at the top of the stack at one of these four places. Coming in from the east, we always ended up at Lambourne. We would be told to 'Enter the hold at Lambourne. Expect 25 minutes delay.' So, we'd slow to the most efficient speed for the altitude and start flying around in circles over Lambourne. If there were 12 aircraft in the stack, we'd start at around 19,000 ft. Aircraft get spat out the bottom at 7,000 ft, and each aircraft in the stack is 1,000 ft higher than the preceding one. Each time an aircraft is spat out the bottom, everyone gets to descend 1,000 ft and more aircraft come in on top. I suspect the concept was designed by someone who was into line dancing.

In its own way, the choreography of it all is quite impressive. For passengers, if you are near a window, you get to check out Lambourne and surrounds and worry about missing your connection. For air traffic control, they get to do a satisfying job coordinating arrivals to an airport from four different stacks. For the pilots, it's just more of the same, maybe another cup of coffee, some slight dizziness from the circling and, if you are on overtime, not the worst thing to have ever happened.

You may remember how we discussed setting our altimeters to standard atmosphere pressure settings on climb after departure. Well, at some point on descent that needs to be reversed so that the correct pressure setting for the arrival airport is used. At London Heathrow that usually happens after you leave Lambourne, heading for the airport. Store that one away for your dinner party grab bag of useless facts.

APPROACH AND LANDING

For pilots, the most interesting phase of the flight is the approach and landing. This is where the adrenaline levels rise, because, like it or not, pilots are judged by their peers and the passengers on how smoothly this all goes. You have to remember that there are very large but fragile egos at play here. If you do a good landing, everyone thinks you are made of the right stuff, and after stopping at the terminal, you will hurry to the front door for the kudos and tips. (I got my best tip ever from an old American. He told me to 'never bet on a slow racehorse'.) If you do a shit landing, everyone on the flight deck goes quiet, and after stopping at the terminal, you send your first officer to the front door to apologise and pay compensation claims.

Pilots of large international jet aircraft might only get one or two landings a month, so if they screw it up, redemption can be a long time coming. Pilots of smaller domestic aircraft can get three landings in a day, so they have no excuse and should be shamed, and maybe even beaten, if they do a bad landing.

Older pilots like to muse about how they will cope with their last ever landing. What happens if it is bad? The chances of this are elevated, too, because of the pressure of having to live out your final years dwelling on it. Some say they will wait until near retirement and then after doing a really good landing, they will let the first officer do the remaining sectors. But how could you not fly your last ever flight? That's a big call.

Thankfully, this wasn't an issue for me, because, thanks to Covid, I didn't know my last flight was my last. It was into Singapore from London, and the landing wasn't my normal high standard, but it wasn't bad either. As a consequence, I don't really give a shit, and my therapist says I'm doing just fine. We should be able to extend to monthly meetings from our current weekly ones as soon as 2025.

Now that we have set the scene for this chapter, it's time to cover a couple of items of pre-approach admin. Have you heard of the 'sterile cockpit'? This is not where you spray the cockpit in petrol and set fire to it to burn all the dried skin and dropped food, although I've seen cockpits where that would have been helpful. A sterile cockpit is one where only items related to the direct operation of the flight are allowed to be discussed. The reason for this is to remove distractions and ensure focus at a time of high workload. Usually, it is instigated at a certain altitude on descent. From that time on, it's not okay to talk about shopping, gardening, arts and crafts and the like, or the other pilots will get grumpy.

Now I know you will have heard about momentum, and large heavy aircraft have a lot of it. In the environment approaching an airport, it's like going down a funnel. As you go, and everything gets closer in proximity, you need to fly more accurately. On large aircraft, you can't turn on a dime, so a bit of anticipation is required. Some say it is like dancing with a fat lady. Fat ladies are just as much fun to dance with, but you need to guide them around the floor and anticipate. Last-minute muscling just won't work. It's called inertia.

The goal of air traffic control is to position your aircraft at the start of the final approach at the right speed, just the right amount of distance behind the preceding aircraft and the same distance ahead of the following aircraft. Once you are in this position, you can use a few different types of electronic guidance to make the aircraft follow the approach down to the runway. These systems are getting more user friendly and accurate as time goes by. In a nutshell, they tell you when you are on the designated lateral track and vertical profile, or your deviation from it. Autopilots are very good at following these profiles. Most pilots acknowledge this fact.

The distance between aircraft on approach can sometimes be an issue if pilots don't stick to standard speeds. Normally, air traffic control is onto this and starts asking pilots to 'rattle their dags' if they are getting too slow or put their foot on the brakes if they are going too fast.

You wouldn't normally expect air traffic control to be the problem, but they were one time in South Africa. We had flown from Sydney during the day and were down on fuel due to the headwinds being higher than forecast. Approaching the Johannesburg Airport, air traffic control gave us a vector that took us out about 20 nm to the north before coming in for a landing to the south. By the time we got to the start of the approach, we were down to 45 minutes of fuel remaining. Just as we commenced the approach, the controller cut a B737 in onto the approach in front of us. I immediately slowed down to minimum approach speed, but we were still gaining on the 737.

My first officer, Tim, said, 'Better be ready for a go-around if he doesn't get off the runway in time.'

I said, 'We aren't going around. If we do, we'll run out of fuel before we get another go. If he's still on the runway, we're touching down behind him.' Luckily, he had just vacated the runway as we entered the flare to land.

Sometimes, when the weather is good, air traffic control will clear you for a 'visual approach'. This means you fly manually, line up with the runway visually and keep a vertical profile, based on flight path angle indicator lights that you can see near the runway on the ground. What could possibly go wrong? I'm glad you asked.

I flew a sector on a 747, 20 years ago, from Singapore to Jakarta at night. The weather was clear and when we turned on to final approach at 3,000 ft, we were cleared for an ILS (instrument landing system) approach onto runway 25 right. The weather was so good that I decided to fly the approach visually and use the ILS as a backup. The approach lights were very bright, so there was no problem flying the approach, but as we passed over them, everything went black. I was already flaring to land, and the 747's landing lights were lighting up the runway markings, so I continued to land. I then used fairly hard braking to bring the jet to a halt halfway down the runway.

The first officer asked the tower controller, 'Is there a problem with the runway lights?' Suddenly the whole place lit up like a Christmas tree. For some reason all the airport lights had gone off at just the wrong time.

Moving on . . . as you start down the approach, a few things always happen at around the same place. As the pilot slows the aircraft down, they extend the wing flaps to increase the camber of the wing and create more lift at lower speeds. At some point the landing gear will be extended. This process normally starts with a low thump followed by a lot of mechanical and airflow noises as the gear falls into place. Think Nirvana playing the chorus of 'Smells Like Teen Spirit' live in a force ten hurricane, and that's your gear extension noise.

I asked an engineer mate about the initial thump, then wished I hadn't. (Feel free to skip this paragraph if you don't give a shit about the initial thump.) I think what he said was 'The hydraulic selector valve is porting hydraulic pressure to the up-locks and actuator. Typically, an electric solenoid valve opens, which directs hydraulic pressure to one side or another of a valve that slides over and opens a port to send the hydraulic pressure to release the up-locks and actuators to power down the gear. The solenoids act very quickly (within milliseconds), and when you have a hydraulic system at 3,000 psi, the valve moves very quickly, and you get a hydraulic "hammer" from the almost instant change in pressure.'

In English, this roughly translates to, 'It's all okay. Nothing to see here.'

Further down the approach, if you are still in cloud, concern will be registering on the pilot's brow. Except in Cat II low visibility procedures, you need to be able to see for landing. Pilots will know from the weather report what altitude to expect to be able to start seeing approach lights and then the runway. If you're in heavy rain, sometimes it can look like you are still in cloud. I turned the wipers on once and was instantly able to see everything I needed.

When the runway is wet, pilots need to have pre-calculated the landing distance required for the conditions. Wet and slippery runways add considerable length to that which is required in dry conditions.

If it is windy, pilots need to be ready for windshear and sudden increases in turbulence near the ground due to the wind passing over the terminal buildings and infrastructure.

When the aircraft is really close to the runway, the pilot or the auto-land will raise the nose to lower the rate of descent and, just before touchdown, reduce the thrust to idle. On the A380 you didn't need to remember to close the thrust levers, because the system would generate aural commands to remind you. How they came up with the words 'retard, retard' is a lingering question. It was hard not to take it personally at times when you were feeling vulnerable.

After the nose raise and the retard, the wheels will then touch down, which is called 'the landing'. There are a few things to consider with the landing.

1. The point at which you start the process differs between aircraft. On an A380 and 747, it will be when the main wheels are about 50 ft above the ground.
2. Some aircraft are harder to land than others. The B767 was an ego killer.
3. If you are good enough, the best landings come from delaying the raising of the nose, but doing it slightly more quickly when you do. If you misjudge this, it will be bad for the ego. You know when you've got it wrong even before you feel it, because you hear all the other pilots sucking air through their teeth and you feel them all bracing.
4. The best landings also involve having a low rate of descent right until touchdown.

5. None of this matters if you aren't a pilot. Non-pilots rescuing aircraft only happens in the movies.
6. Aircraft successfully landing upside down also only happens in the movies, aided by cocaine and booze.

As previously mentioned, there's a lot of ego involved with the landing. Some pilots seem to be very good at it and rarely do a bad landing. Others seem to have no clue. It's a fact that if you continuously did bad landings, the airline might recommend you for a promotion to 'civilian', as bad landings scare passengers.

In a really bad landing, the three million plus parts in a large aircraft all groan in unison as they try to separate from each other. Some landings have been known to cause the overhead oxygen masks to fall down from the ceiling. This is known as the 'rubber jungle'. In this case, it's funny to see how many people are stupid enough to put on an oxygen mask when they know with some certainty that they are on the ground.

The worst landing I ever experienced was when I was a first officer. We had to fly from Sydney to Christchurch and back during the day. The captain flew to Christchurch and thumped the aircraft onto the runway so that there was no doubt that we had arrived. At the terminal he said to me, 'If you don't mind, I'll fly back as well so I get these landings sorted.'

Going back into Sydney, it was now dark, and this time we landed with enough force to bring the dinosaurs back to life. The boss achieved a rare rubber jungle, a complete loss of mojo, and, as for me, the force of the deceleration gave me an instant headache. Neither of us went out to face the passengers that night. We were too scared to open the cockpit door.

Crosswinds complicate things on landings. They happen as a

result of the wind direction not being aligned with the runway. There's a lot of skill involved, which is why pilots nearly always practise with crosswind in the simulator. A well-known fact is that when airports are designed, things like the prevailing wind direction and the position of politicians' houses are both taken into account. For these reasons crosswind landings are still quite common.

I've given you a lot to think about already, but supposing all of this has happened without issue, as it does 650 times a day at London Heathrow, then the aircraft just needs to be brought down to taxi speed before the end of the runway by using a combination of wing spoilers, reverse thrust and wheel braking. It's as simple as that.

Sometimes, however, a pilot will not be happy with the landing conditions and will decide to abort the landing, which is known as a go-around. This can be for a whole lot of reasons, including not being able to see the runway due to cloud or rain, windshear on the final approach, another aircraft – or people or animals – on the runway, and equipment failure that needs to be sorted before landing.

A go-around can be a bit of shock for passengers, but they should take heart from the fact that the pilot is being conservative. In a go-around, initially the nose rises up quite rapidly and the thrust increases as well. There will be all the noises associated with retracting the landing gear (the same song played backwards). It all seems a bit frantic . . . because it is. You probably won't hear from the pilot for a bit either, as they will be busy getting ready for the next approach.

I did about ten go-arounds in my whole career, so they are not that common, although we pilots are always mentally prepared for one and well aware of how it affects passengers.

On that note, I'm going to try and put a positive spin on the missed approach experience, to show you just how good it can be. Here goes. Passengers who find flying alarming on a good day will be even more alarmed during a go-around. If alarmed is what you are after, then this is your gig. Passengers who are not happy unless they are anxious will end up much happier. Passengers who love flying will get more bang for their buck. Passengers with tight connections will swear a bit and look for anyone else to blame. Finding someone will justify their anger. Passengers who don't really give a shit will still not give a shit or possibly even reduce their level of shit giving. Airline ground staff will run around in circles because the next flight will be delayed, and hence they will get some much-needed exercise. Pilots will get more flying practice and more overtime. Airport car park owners will rub their hands together at the extra revenue. Conspiracy theorists will blame it on climate change. Media will be happy because it gives them some drama to report on, and they can pull out the stock photo of the wrong aircraft type with flames coming from the back of an engine. The dog on the runway who caused it will have had a ball chasing away birds and will be rewarded with a pat and a feed from airport authorities, then driven home in a van.

It seems that there are a lot of winners and few losers from a go-around, so perhaps more should be planned.

AFTER LANDING

As we taxi off the runway, it's effectively the beginning of the end. This is the point where you need to reactivate your wits and shake off that last drink, because it's the beginning of your own personal Super Mario game. In the game, you try to reach your safe place while everything possible is thrown at you to impede your progress.

As the aircraft exits the runway, the pilots start 'cleaning it up' by bringing up the wing flaps and starting the auxiliary power unit. The cabin crew will start a tiresome waffle welcoming you to such and such and so on and so forth. They will conclude with, 'Please remain in your seats until we are safely stopped at the terminal,' but there is always one person who thinks this doesn't apply to them. The intense silent loathing directed at said passenger could be used as a weapon in a war, much like a laser beam.

Murphy travels on every flight, including yours. If you are 45 minutes early, the aircraft currently occupying your bay won't have left yet, so all that hard work in arriving early is for naught. If you are 45 minutes late, the airport will have sold your bay to another airline, so you will be parked on a remote bay and have to catch a bus to the terminal. This will make you later still.

Once you are at the gate (or bus) and the engines are shutdown, you will often be able to hear dogs barking. You are not hallucinating. There are frequently dogs in the cargo compartment underneath the cabin, and they get just as excited about arriving as you. They brag to each other about what their owner has planned for them, the fact that they get to sleep on the owner's bed and the size of the sticks they are going to chase when they get home.

When the aircraft stops at the bay and the seatbelt sign is turned off, this is the cue for all 500 people to stand up at once. I worry that this will throw the Earth off its axis, but so far it hasn't. Normally the passengers at the front get off first, so they are in no hurry getting their shit together. You can see this from down the back. Initially the pace of disembarking is glacial, and it slows up after that. Often, people seem surprised when they can finally move, so they aren't ready. This slows things down even more to the point where you can hear the slowness.

Finally, you will get out the door to find that you were there so long, there's been a change of government, and cash, cars and commonsense have been outlawed. Now starts the 'running of the bulls' through the terminal, which is similar to the Pamplona version but more dangerous. In combination with our Benny Hill chase mentality, it can be quite exciting, trying to get in front of as many people as possible before the customs and immigration queues.

Sometimes customs and immigration will be empty when you arrive. When this happens to me, I look for hidden cameras, as I think I must be on some 'set-up' comedy TV show (it was

Candid Camera in my day). Other times you'll arrive to find that the queues reach past where you got off your flight, and you have to backtrack to find the end of them. This scenario is more likely by a ratio of 100:1.

Later that day, you will get to the front of the queue and be told you are in the wrong one. Most of the time, the officers have a heart and deal with you anyway – unless you are a foreigner.

So now, having taken half a day to reach the baggage claim area, you would be correct in assuming your bag would be waiting for you. I hope it is, because no one deserves to have their bag randomly chosen for a trip to Saturn . . . Wait. That's not entirely true. I know people who deserve it, but I also know my editor will object if I print them here, so I won't.

If your bag is not waiting for you and you've been on the ground for more than an hour, it's time to go and enquire at the lost bags counter. The person behind the counter will look stressed and weary. They will have been dealing with nothing but angry, frustrated people who are tired and just want to get out of there. That must be the best job in the world, and something I could see myself doing part-time for some fun. 'Yes, sir. I can confirm your bag is missing as it says so here on our computer. It says it's on its way to Karachi. I would have told you an hour ago on the airport PA, but I had to take a coffee break and then I forgot.'

One evil trick you can pull on these lost baggage people is to be nice to them. That will confuse the bejesus out of them and probably streamline their entry into their inevitable therapy.

If your bag is missing, there is no guarantee of it ever turning up again. You can't use logic here. It doesn't apply. You just have to roll with the punches. Legal action won't help either. The last man who took an airline to court over lost luggage lost his case.

SEVEN

THE FUTURE OF FLYING

A LOOK INTO MADAM ZELDA'S CRYSTAL BALL.

WHY AVIATION IS SAFE

It's a fact that aviation is one of the safest ways to travel. This didn't just happen on its own. There have been many steps taken over the last 80 years to improve aviation safety.

The establishment of the International Civil Aviation Organization (ICAO) in 1944 marked a significant milestone in adopting a global approach to aviation safety. Through the creation of standards and regulations, ICAO has defined the rules that make aviation one of the safest modes of transportation.

Agencies like the Federal Aviation Administration (FAA) in the USA, and their counterparts in other countries, play a crucial role in creating and enforcing rules that ensure our safety. They oversee various aspects, including pilot regulations, aircraft certification, maintenance requirements, air traffic control, airport operations and other critical safety considerations. Additionally, they facilitate information sharing among countries and continents, promoting transparency and enabling the identification of any non-compliance with international standards.

Civil aviation regulators in each country decide when to ban another country's airlines from their territory, due to the latter country's shonky operation. For example, Nepal Airlines currently aren't allowed to operate into Europe. Not only does this reduce the chances of that airline killing nationals of Europe, but it means that the local civil aviation mob doesn't have to deal with any 'clean-ups' when accidents occur.

Even approved airlines are subject to spot checks by civil aviation authorities. These can happen after landing or before take-off. The authorities have the right to enter a foreign aircraft and demand to see all licences, paperwork and approvals as well as required equipment on board. Non-cooperation or failure could mean the aircraft being grounded indefinitely.

A regulatory check happened to me before a 747 departure in Frankfurt a few years back. We were heading to Singapore and had a crew of four pilots. Three German officials came on board sporting serious looks and official IDs. The leader looked like Major Hochstetter from *Hogan's Heroes*, but although he was demanding to see our 'papers', he didn't say it in that way. He was pleasant enough, too, which was a relief.

We had just got onboard, and I quickly realised that there was no sign of Andrew, one of the second officers. Then I noticed him peering through a crack in the curtain of the crew rest and beckoning to me. I went over and asked what the problem was. He looked very flustered. He told me that he didn't have his pilot's licence with him, which is an instant failure. If the officials were to find out, and they would, it would mean he would be stood down from the crew and flown home as a passenger. This is not career ending but it's a major blip in your career, costing you many dollars. If pulling you off the flight meant the flight couldn't go, the major blip would go up a few levels to a three-star general blip.

I told Andrew to hide behind the chair in the crew rest with the lights off, and I pulled the curtain. The officials checked us three remaining pilots' licences and eventually left. They had no idea that there were four of us. After the officials left, we three hauled Andrew out of the crew rest. It was free beer for all three of us in Singapore, and it did taste very good.

We've talked about how good pilot training is these days with state-of-the-art simulators and crew resource management training. There will always be airline management knobs who try to downplay the influence good pilots have on an operation. Calls for single pilot operations have risen lately, pushed by people with a cost cutting agenda. There can be no other reason. These people have a stake in cutting costs to raise bonuses and share-holder profits. They trot out the same tired arguments and throw the dice on whether they will still be around when something goes wrong as a result of their greed.

Just culture is another recent development that has enhanced safety. Just culture is a concept which emphasises that mistakes are generally a product of faulty organisational culture, so it focuses on discovering what went wrong rather than simply punishing individuals for mistakes. It's important to note that a just culture is not a no-blame culture, and individuals will still be accountable for any wilful misconduct or gross negligence.

Technology advances in the cockpit have had a lot to do with the increase in aviation safety. Modern air traffic control systems have also improved overtime to enhance safety. They will continue to get better, as will aircraft construction. But the technology in the cockpit is something that just keeps improving all the time.

The traffic collision avoidance system (TCAS) was a huge advance in safety. This system allows all aircraft fitted with a

TCAS transponder to electronically find out about the position and altitude of nearby aircraft. This is kind of like each aircraft sitting down with another aircraft for a cup of tea and saying, 'Oh hello. What height are you at and what direction are you travelling in?' Then the second aircraft will say, 'I'm at 35,000 ft on a heading of 187. How about you?' But it actually happens a bit quicker than that. This interrogation-and-response cycle may occur several times per second.

The TCAS then creates a three-dimensional map of all nearby aircraft and calculates the risk of any collisions. For any aircraft that are likely heading for a crash, the TCAS will negotiate an evasive manoeuvre. Now, this is where the cup of tea gets thrown away as there are more important things to discuss. The first aircraft will say, 'Oh dear. It looks like we might crash into each other if we don't do something, and I'm not keen on that, so how about I climb 500 ft and you descend 500 ft, and reasonably quickly would be good.' The second aircraft then replies with, 'Sterling idea, old chap. Consider it done.' The TCAS then gives these instructions to the flight crew on a cockpit display and using a synthesised voice. In this way, mid-air collisions have almost become a thing of the past.

The onboard terrain awareness and warning system (TAWS) has also revolutionised safety. It's aimed at preventing unintentional impacts with the ground, using a worldwide database of terrain and comparing it to the aircraft's actual position. If TAWS decides there's a danger of running into a hill, it warns the pilot with aural warnings and a visual display on their navigation system.

When I was a young navigator in the Royal New Zealand Air Force, I was called out to navigate a search and rescue aircraft on 28 November 1979. The search was for a missing Air New Zealand DC10. Sadly, the aircraft was destroyed on the side of Mount Erebus with the loss of 257 people. The reason the accident happened was that the navigation coordinates had been changed the night before without telling the pilots. The pilots thought they were flying down McMurdo Sound, but they were in fact flying straight into the side of Mount Erebus. TAWS would have stopped this accident from happening.

So, as you can see, things just keep getting better in the safety world, and let's hope they continue to do so. But what does the future of aviation look like? Let's have a gaze into the crystal ball.

WHERE TO FROM HERE?

What does the future hold for aviation? Well, sometimes history is a good indicator for the future. So how have things changed in the 47 years since I started flying in 1977?

Aircraft have become more efficient in terms of fuel per passenger mile, and airfares have relatively reduced in price. We've been through and are coming out of the 'four-engine very large aircraft' phase, although you could argue that it's not the size that's changed, but the preference for two engines has increased. Indeed, the new two-engine B777X has nearly the carrying capacity of the 747. Available engine thrust has more than doubled since 1977 from around 50,000 lb to well over 100,000 lb per engine. Aircraft are now lighter thanks to composites, and more efficient and quieter, thanks to advances in aerodynamics.

Computers rule now. Fly by wire is the thing. It's become much harder to kill yourself thanks to safety systems built into the controls on aircraft. Other systems such as windshear warnings, weather radars, ground proximity warnings and collision avoidance systems have enhanced safety.

On the flight deck, crew resource management has made crews more effective. Paper has almost been eliminated with the use of iPads and electronic uploads of flight plans and weather. Communications have improved, with ACARS operating through satellites and VHF. HF still exists in certain areas, but will soon be gone. Training systems are better. The use of the autopilot systems to allow pilots to concentrate on important management matters is mandatory, although manual manipulation skills have suffered a little as a result.

The romance of aviation has gone, thanks to the accountants and the terrorists. Flying has become something to suffer through to get somewhere. It's become a commodity. Airports are large shopping centres. Passengers are made to feel guilty for their carbon footprint by grumpy Greens.

Low-cost airlines keep popping up and disappearing like Whac-A-Moles. The ones that survive are generally subsidised by parent airlines or are bottom-of-the-barrel operations that get by using slave labour. Let's face it. What is the difference between full service and low cost? The aircraft cost the same, as does the fuel. Passengers don't get food and movies, and I very much doubt if the aircraft are cleaned as often. The low-cost part is because they work their staff harder for less money. This formula doesn't necessarily attract the highest quality people.

If you buy a ticket on a low-cost carrier, it's a lottery. If they decide it's going to cost them money to run the flight, they'll probably cancel it, even though that's illegal. (Some premium airlines are guilty of this as well.) All they need to do is blame crewing or maintenance, and they'll get away with it. When they cancel your flight at the last minute, don't expect any compensation, and you'd better be quick to grab the nearest bus stop seat, because they won't be paying for a hotel room either. With low-cost airlines, bankruptcy is always imminent. If you buy a ticket on one, it's a gamble.

Despite this, I have to come clean and admit to travelling on low cost carriers when it's been convenient. Normally when you front up to the counter, they will ask you if you have any reservations. This is where you quickly say, 'Yes. Many. But I'll fly with you anyway.'

From a pilot's perspective, pay rates have dropped relative to other professional occupations, and pilots generally work harder. Work-life balance is challenging, particularly for international pilots. That said, it's still not bad, except for the paper coffee cups.

On the non-flying side of companies, security departments have gotten huge, HR departments have become a thing, as have change departments. People generally don't wear ties anymore, and affirmative action in employment is commonplace. Management self-enrichment has become a sport.

So what does the future hold? Well, my crystal ball is old, but I'll have a good go at it without thinking outside the square too much. The long aftermath of Covid will be difficult for airlines to navigate. I don't even wish the current situation on people like Alan Joyce. (Actually, that's not true.) It was hard enough for airlines to make a profit when things were going well.

The last 30-year macros of lowering interest rates and the massive increase in global debt have left the financial system at risk of a colossal collapse. This mountain of debt has allowed people to live their lives far into the future with travel and asset purchases. Unfortunately, piling on cheap debt to keep the house of cards from collapsing has made it worse. A re-balance might take ten years or longer to shake out the imbalances.

At the same time, technological advances won't stop. Does this mean pilots will become superfluous? Are modern safety systems making us glorified bus drivers? Single pilot flight decks might become appropriate at some point, but do you really want to be a passenger on an aircraft with no pilots when the shit hits the fan, as it did on QF32 out of Singapore in 2010? They were lucky to have four experienced pilots in the cockpit that day, and it was still a massive effort to save the aircraft. And it was only the skill and experience of Captain Kevin Sullivan that saved QF72 north of Perth in October 2008.

The use of AI (artificial intelligence) will definitely enhance safety. Air traffic control systems will take over a flight from airborne to landing, with pilots and controllers just monitoring. Better air traffic control systems will allow more efficient routing, thus cutting costs. In addition, engineers will be able to fix most problems remotely. Without a doubt, innovations in aviation will come that we can't even imagine these days.

There is talk of bringing back supersonic aircraft, and as long as noise problems can be overcome, then why not? But this will still be an expensive way to travel. They say large four-engine aircraft are dead, but there will eventually be two-engine aircraft much bigger than the A380 of today. There has to be, or there won't be enough room in the skies for all the aircraft. In an ideal world, the world's population would stop growing, which would relieve some of these growth pressures.

Data mining by airlines will increasingly become important. This is where companies use large amounts of data to find anomalies, patterns and correlations, which can be extrapolated to predict future events. They can use these predictions to increase income and improve customer service while reducing costs and risks.

Efficiency will be paramount. With the demand for action on climate change, sustainable fuels will need to become commonplace. Small electric aircraft are already in use, but these are currently limited in size and range. It is thought that hybrid aircraft may be the answer, using hydrogen to run an engine that creates electric power for electric motors for the flight. Aircraft shape is likely to change as well, with shapes like 'flying wings' being less noisy and more efficient.

Last but not least, more efficient ground-based transportation needs to be considered as well. Japan's Maglev train has just hit 600 kph. That means it would be capable of going from Sydney to Melbourne in 90 minutes or Auckland to Wellington in an hour. Given that this means cutting out two 30-minute taxi rides from airports to cities, that's way faster than flying, and it would be my choice. This, of course raises the possibility of 'train lag' or 'mag lag', and you can say you heard that here first. Domestic aviation will be dead between those points once this happens and populations get large enough to support a half-hourly service. It bears thinking about.

GREEN FLYING

Here's something to contemplate. Should you be feeling guilty about your carbon footprint while you fly? Does Greta Thunberg have a case for frowning at you or should she be out the back with the celebrity chefs?

A common theme in aviation is that all the fun has gone out of it. The terrorists ruined it with 9/11, and now everyone wants you to feel guilty because of your carbon footprint. From the crew's point of view, management have lost their sense of humour too. The job needs to be fun, because happy workers make for happy customers. But you can't have a joke anymore because someone will get offended, or do anything out of the ordinary because the safety department will get grumpy. A lot of the anecdotes in this book come from a less serious time when people were realistic, and fun was encouraged.

But what of your carbon footprint? Let's look at some facts. Aviation's contribution to climate change is about 3.5 per cent of warming, or 2.5 per cent of CO_2 emissions, which, contrary to popular opinion, makes its contribution to emissions relatively small on an industry comparison.

Passenger cars and large trucks are by far greater polluters, because there are far more of them. Many studies have been done on the relative carbon footprint of driving a car versus flying the same distance. On a per kilometre basis, the air carbon footprint per passenger kilometre is slightly lower than driving. The longer the flight, the better it gets, as more fuel is used in the take-off phase than the cruise phase. It should be noted, however, that these figures are based on one passenger in the car, which is not always the case. Sixteen Taliban fighters all piled into one Toyota HiLux wins this contest hands down, and I'm always impressed by their commitment to fighting climate change. Maybe they should be nominated for a Nobel Peace Prize too. But under normal circumstances, the case in favour of air travel increases because it is direct, whereas car travel follows roads, which are notorious for going the long way around obstacles.

Reducing the carbon footprint of aviation is no easy task. Sustainable fuels make up only a small percentage of all aviation fuel at the moment, and scaling this up will take a concerted effort over time, given the current limits on blending sustainable fuels with standard jet fuel.

There are some promising design concepts emerging. Airbus, for example, are planning to have the first zero-emissions aircraft by 2035, using hydrogen fuel cells. If need be, it will achieve this by being kept in a hangar and never used, but it will still sound impressive. Electric planes may eventually become feasible, but the current battery weight and power limitations are likely to limit them to being very small.

So, as you can see, flying green is not something that will happen in a hurry or that you can do much about. At least you can now say you have thought about it. So, I recommend having another drink and enjoying your movie.

THE FINAL WORD

For most airline workers, including me, working in the industry is and was a privilege. I saw the world, and I met wonderful people both on the aircraft and off. I wouldn't change a thing and I'd probably do it again too. I might sound a bit cynical about a few things in aviation, but in general, it's been a blast. I hope this book has helped you understand aviation better. It really is the most amazing thing.

ACKNOWLEDGEMENTS

I'd like to acknowledge all published and aspiring authors every-where. Reading has changed my life for the better. Without your example, I wouldn't have a clue how to write either.

Thanks to my agent, Tom Gilliatt, for having faith in me and not giving up.

Thanks to Pan Macmillan, particularly Alex Lloyd, Belinda Huang and Nikki Lusk, for your subtle advice and enduring patience.

Thanks to my dad for setting an example of a really good man, and my mum for demanding excellence in everything I did. Thanks to my sons, Jamie, Dean and Doug, for giving me a reason to be inspirational.

To the crew members who contributed to the parts of the book that put certain people under the spotlight, your bravery is very much appreciated and your names have been 'forgotten' already.

To all the pilots and cabin crew I ever interacted with in my

37-year aviation career, you are all stars. Your pursuit of excellence, sometimes under the most trying conditions, is unwavering.

And to Winston Churchill, who is most quoted about never giving up. You were right, sir!